In the climate action camp we so often h
"mitigation" box – whether at workshops
Adaptations this unhelpful dichotomy is
powerful real life examples. Morgan's sto
and hopeful. It serves as a huge challeng
"humanely" – not just "humanly" – possible.
**Sonja Graham, CEO Global Action Plan**

The writing may be on the wall about the speed and scale of the rapid transition needed to cut pollution and adapt to an increasingly volatile climate, but it takes someone truly able to *see* to be able to read it. Morgan Phillips has accomplished that admirably in Great Adaptations. He vividly brings to life the immediate threats and opportunities for how we must change to all thrive within the boundaries of the biosphere. Here is a twin wind of reality and hope.

**Andrew Simms, Rapid Transition Alliance**

It is only by embracing adaptation that will we be able to build the solidarity and belonging that can bring us out of the crisis. And, as Great Adaptations shows, it has to be adaptation from below, not passive feed-down from techno-fixes and bureaucratic intervention.

**Dr. Wolfgang Knorr, Lund University**

As peaceful warriors, conscientious protectors, activists, rebels and change makers the question now is how do we become our most adaptive and abundant collective selves? Great Adaptations is a precious offering of incisive pathways into conversations that might liberate us from underneath the heavy weight of 'fixing' the world to place us into the bright dawn of creating the world of our longings. Read it and leap!

**Skeena Rathor, Co-Founder, Extinction Rebellion**

My earnest hope is that this book will be a turning of the tide; and that, with the silence broken, the world can finally begin the painful process of awakening properly to climate reality... including to the reality of how we must now adapt transformatively, if we are to have any chance of heading off eco-induced collapses.

**Prof. Rupert Read, University of East Anglia**

It is becoming ever clearer that the world system is in deep trouble, above all the climate crisis. Great Adaptations highlights how we not only need to engage with adaptation as a neglected sub-topic of climate change, but how we need to examine the role it plays in either cementing existing inequalities, or breaking them down. Great Adaptations is provocative, informative and imaginative.

**Carne Ross, Founder, Independent Diplomat**

We stand on the edge of climate catastrophe that threatens billions around the world and is unpicking the very life systems we rely on. With the time for tinkering around the edges long past, what is needed now is bold and visionary thinking - this is an important contribution to imagining a different future.
**Asad Rehman, Executive Director, War on Want**

Great Adaptations is a revelation in terms of unravelling novel conversations and contentions on the topic of climate adaptation and is a remarkable attempt to forge a mutually reinforcing relation with climate mitigation. It forays into diverse sub-topics within the broader spectrum, including topics such as deep adaptation, as well as an introspection of maladaptive strategies, framing arguments on the need to channel more climate finance into adaptation projects and reiteration of the importance of creating a befitting conception of adaptation within the public minds. The author has managed to pen an intriguing combination of confluences and dichotomies, thus creating pathways for further contemplation.
**Prof. Dr. Saleemul Huq, Director, ICCAD; Chair, Expert Advisory Group, CVF**

This is a brave and timely work, one that not only seeks to raise the profile of adaptation, in a world already experiencing the impacts of 1.2°C of warming, but one that also seeks a greater goal, beyond the current obsession with technical fixes and towards the real heart of the issue. In challenging the prevailing politico-economic paradigm that shapes human development as a fight and a competition, Morgan seeks to frame adaptation in the light of social justice and plots a path towards a more co-operative and kinder future.
**Richard McIlwain, Deputy CEO, Keep Britain Tidy**

I know Morgan to be an impressive thinker – someone who follows where the climate science leads, rather than bending to what is politically feasible. Someone who understands that any proportionate response to climate change must flow from searching questions about who we are and what we value.
**Dr. Tom Crompton, Director, Common Cause Foundation**

In 200+ episodes, recorded over six and a half years, not once have we talked meaningfully about climate adaptation. Morgan is too polite to put it this way, but his powerful book asks "why the hell not?". Great Adaptations is a wake up call to gobby podcasters, and - more importantly - to a sector that, in the rich world at least, talks lazily of "fixing" or "stopping" climate change while the fires burn hotter and the hurricanes blow harder.
**Ollie Hayes, co-host Sustainababble podcast**

# Great Adaptations

## In the shadow of a climate crisis

**Morgan Phillips**
The Glacier Trust

Great Adaptations
By Morgan Phillips
© Morgan Phillips, The Glacier Trust
ISBN: 9781912092147

First published in 2021 by Arkbound Foundation (Publishers)

Cover image by Suzie Harrison
Design by Hannah Ahmed

No part of this publication may be reproduced, stored in a retrieval system, or transmitted, in any form or by any means without the prior permission of the publisher, nor be otherwise circulated in any form of binding or cover other than that in which it is published and without a similar condition being imposed on the subsequent purchaser.

Arkbound is a social enterprise that aims to promote social inclusion, community development and artistic talent. It sponsors publications by disadvantaged authors and covers issues that engage wider social concerns. Arkbound fully embraces sustainability and environmental protection.

This book is printed by Calverts Ltd., a workers' cooperative, on pre-consumer recycled paper, using vegetable oil-basedinks, and electricity from renewable energy sources.

Arkbound
4 Rogart Street
Glasgow G40 2AA
www.arkbound.com

Dedicated to:

**Surbir Sthapit**
1967 – 2020


**Dilly Phillips**
1949 – 2021

# Acknowledgements

This book has been a team effort. I have been supported by many brilliant people. Thank you to Andy Rutherford, trustee of The Glacier Trust, who has read and commented on many drafts of this book; he has been my editor, critical friend, and coordinator of a wider group of reviewers who I will list below. Thank you to Hannah Ahmed, who has provided so much thought, creativity and talent to this project. Hannah has designed this book and given it its beauty. I wanted this book to be a tangible, shareable object, she has made that possible with great humour and patience. Thank you to Suzie Harrison for the papercut illustrations that adorn the front cover and each part of this book; thank you too, for donating the original artworks to The Glacier Trust. Thank you to Steve McNaught and the team at Arkbound for the immense amount of work that has gone into the publication of this book. Thank you to Ellen Tully for volunteering with The Glacier Trust and for providing incredible support in sourcing images, editing chapters and designing assets for our social media campaign. Thank you to my publicist Elly Donavon who has guided me through the promotional work, helping the book and its messages to reach the audiences we wanted it to find.

Thank you to everyone who commented on drafts of the book. Any errors within it are mine and mine alone; and as reviewers you are not responsible for any of the opinions or inaccuracies the book contains; my sincere thanks go to Andy Hillier, Ann Phillips, Carys Richards, Craig Hutton, Dinanath Bhandari, Ellen Tully, Kim Dowsett, Marcela Terán, Mary Peart, Richard

Phillips, and Rupert Read.

Thank you to everyone who has pointed me in the direction of inspiring stories and theories, those who have allowed me to use their images, and to everyone who has kept me motivated to start and finish this book. Thank you, Jamie Forsyth, Peter Osborne, Richard Allen, Levison Wood, Robbie Udberg, Emma McQuillan, Ellen Winfield, Glyn Phillips, Elsa Davies, Ceri Jones, Liz Koslov, Amanda Collins, Alice Bell, Andrew Simms, Matt Rendell, Jonny Cave, Daniel Stone, Mark Phillips, Fergal Byrne, Lisa Schipper, Narayan Dhakhal, Krishna Ghirme and all my friends and colleagues at Global Action Plan in the UK, and at EcoHimal, HICODEF, Tribhuvan University, Deusa AFRC, and Mandan Deupur AFRC in Nepal.

I would also like to thank all those I have listed in Chapter 11; they have all inspired me and made me think, as have so many of the other authors, filmmakers, campaigners and academics who I have cited. Finally, thank you to the team at Calverts, everyone who supported the pre-release Crowdfunder, and to the Kenneth Miller Foundation and Margaret Hayman Charitable Trust who support The Glacier Trust's work and the Great Adaptations project specifically.

# Contents

**Introduction** — 03
The 'best-case' scenario is bad enough — 04
Two working assumptions — 09
This is a justice issue — 10
Emerging from the shadows — 16

**Part 1 - Silence** — 20
**1. The storm before the calm** — 24
**2. Code of silence?** — 35

**Part 2 - Adaptation** — 38
**3. Five reasons to talk about adaptation** — 42
**4. Air condition everything** — 46
Welcome to Glasgow — 46
You got Coolth? — 51
Air-conditioned pavements — 57
Generation 22°C — 59
**5. Snow, Grapes, Guns and Dams** — 67
Don't eat the artificial snow — 68
Grape adaptations — 72
Guns don't kill people, climate changes do — 78
Dam good idea? — 82
Adaptation 'first — 90
**6. Adaptation in the wild** — 95
This ice sheet ain't big enough for the both of us — 96
Camels in the crosshairs — 97
Icequake! — 98

| | |
|---|---|
| Small is beautiful...and adaptable | 99 |
| Adder-daptation | 99 |
| Plastic coral | 100 |
| **7. Survival of the friendliest** | 102 |
| Drinking fog | 104 |
| So you've declared a climate emergency? | 109 |
| It's a long way from Warwick to Deusa | 114 |
| Migrate adaptations | 131 |
| **Part 3 - Transformation** | 136 |
| **8. 'It looks bleak. Big deal, it looks bleak.'** | 140 |
| Two degrees of separation | 143 |
| Transformation, collapse, or total collapse? | 150 |
| Whack-a-mole | 158 |
| **9. Different or better kinds of future** | 165 |
| Are collapse events necessary precursors to successor civilisations? | 172 |
| **10. A note from the author** | 178 |
| **Part 4 - Stories** | 182 |
| **11. The Reassuring Story** | 186 |
| The reassuring story (part 2) | 196 |
| **12. Adaptation is unavoidable, but maladaptation is not** | 200 |
| Postscript | 204 |

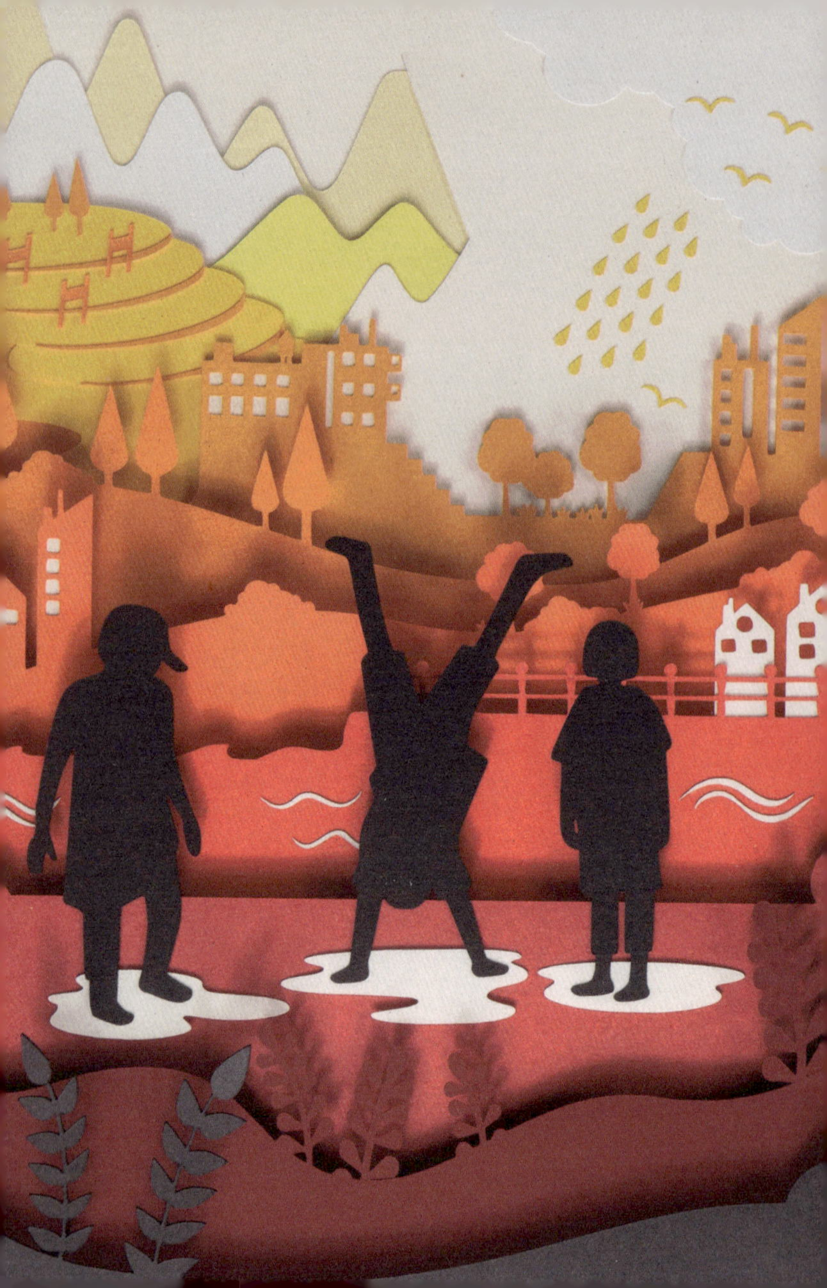

# Great Adaptations

In the shadow of a climate crisis

**Morgan Phillips**
The Glacier Trust

## GREAT ADAPTATIONS

I began writing this book in late 2019, when COVID-19 was something that was happening in a faraway Chinese city and the United Kingdom was pre-occupied by Brexit and a snap General Election. Climate change was on the news agenda, but low on it. This was – *hopefully* – about to change. The UK was preparing to welcome 'COP26', the UN's next big climate conference; the circus was coming to town. By November 2020, the eyes of the world would be on host city Glasgow. Or so we thought.

It is now Spring 2021 and there are no words to adequately describe the impact COVID-19 has had; utterly devastating is the best I can do. We lost a dear member of the Phillips family and a close colleague of The Glacier Trust; this book is dedicated to them both.

Tragically, the shockingly high death toll is still rising, the systems that were supposed to protect the most vulnerable creaked and failed. The machinery of the State has proven to be anything but well adapted to a crisis it was warned about. Amidst all this comes COP26; it has reappeared on the horizon. It is hard to predict what might emerge from it. However, with no US election scheduled, a COVID-19 vaccine programme in place, and a Brexit that is 'done' (if not dusted), there is a chance that climate change might creep a few items further up the news agenda than it would have done in 2020. Delays aren't often a good thing, but this one might work out for the best. We'll see.

And so, with the world in a state of triage, recovery, and flux, it feels like a timely moment to release a book on adapting to a crisis.

# Introduction

I am the UK Co-Director and only paid employee of The Glacier Trust (TGT), an NGO that enables climate change adaptation in Nepal. It is a role I've been in since 2016 and it has opened my eyes more than I ever expected. The things I have witnessed – the seen and unseen impacts of climate change – have been extraordinary. Lives in the Himalayan villages I have visited are on a knife edge. Landslides, floods, glacial retreat, drought, fire, air pollution, and insect pests are haunting the future of an already fragile country; it is on the brink of being turned upside down. But Nepal is only one epicentre of the disaster that is unfolding, what I have learned about the damage the climate crisis is doing to wholly innocent people on every continent is horrifying. I knew that climate change needed to be mitigated, but the need to adapt to it is far greater than I'd ever imagined.

Over the last five years – in my role as TGT's fundraiser, storyteller, director and everything in between – what has struck me is the scale of the damage climate breakdown will do over the coming decades; it is an astonishing amount. It has therefore shocked me how little attention is being given to adaptation as a topic within the environmental movement.

Stories about adaptation are rare. I have been an environmentalist for twenty years and I readily admit to having spent most of those years in almost complete ignorance about adaptation. It wasn't something that came across my radar in my pre-TGT days. It is easily possible to go a whole year as an employee of an environmental NGO and not hear the term 'climate change adaptation' said once. As a consequence, a lot of people, both within and beyond the environmental movement, are still very much in the dark about it. This worries me greatly.

It has thus become something of a personal and organisational mission to bring adaptation out of the shadows. We want to amplify its importance in the minds and work of our colleagues in the environmental movement, and in the wider public. So far, in pursuit of this goal, TGT has (a) published two *'We Need To Talk About Adaptation'* reports to highlight how few mentions adaptation is getting in content produced by the UK's five largest environmental organisations[1]; (b) released a film called *'Coffee. Climate. Community.'* which tells the story of coffee growing as an adaptation strategy in Solukhumbu, Nepal[2]; and (c) contributed a chapter to *'Climate Adaptation: Accounts of Resilience, Self-Sufficiency and Systems Change'* (also published by Arkbound)[3].

*Great Adaptations* is our latest and most ambitious contribution to the adaptation advocacy effort. It is a book, podcast and awareness raising campaign. We touch on how adaptation is talked about and what that means for how it is done, but our primary aim is to provoke discussion and debate. We want to get people talking about adaptation. If *Great Adaptations* sparks one conversation in one house, one workplace, or on one social media feed, it has done what it set out to do.

## The 'best-case' scenario is bad enough

The UK stands poised to welcome world leaders to the twenty sixth UN climate change Conference of the Parties, 'COP26'. It is much anticipated and, as is often the case in the run up to a major 'COP', the mood within the climate movement is oscillating from buoyant optimism to deep despair. The last major COP, number 21, held in Paris in 2015, was judged a success (at least by those whose reputations rested on it being a success). It can't really be seen that way now. Since Paris, the 196 nations who agreed to *'limit global warming to well below 2°C and pursue efforts to limit it to 1.5°C'*[4] have set out how they plan to contribute to this goal. In short, sadly, they're not contributing; at least not very well. Their *planned* contributions don't add up to enough, and their *actual* contributions add up to even less[5]. As Professor Rupert Read puts it: *'Paris achieved what was politically possible, not what is needed.'*[6]

So, this is where things currently stand. World leaders have agreed to limit warming to *well below* 2°C (above pre-industrial levels) but – barring a Glasgow-inspired miracle – this looks unlikely to happen. The Paris agreement is too flimsy, too lenient, and too tightly bound to economic paradigms and political ideologies that are incompatible with the goals being set. There is, however, still a glimmer of hope, the blockers are *only* political. From a purely scientific perspective, stabilisation of global warming at or below 2°C is still possible. Indeed, 1.5°C is still (theoretically) possible and, if mitigation efforts expand and accelerate rapidly enough, there is also a chance that – after stabilisation – a gradual temperature decline might start to kick in. It's all still doable, but I can't emphasise enough how revolutionary the shift in political and economic thinking, and *doing*, needs to be. If current trends persist, there is only a 5% chance that temperature rises will be limited to 2°C. Even if the signatories of the Paris Agreement make good on their pledges, the odds only improve to a 1 in 4 shot - a 26% chance[7]. Things feel dangerously off track.

Remarkably, in spite of the scale of social, cultural, political and economic change required to 'save' the climate, very few have given up on that mission; I haven't. In fact, the urgency and severity of the situation seems to be attracting more people than ever to the cause. The determination to cut greenhouse gas emissions remains admirably strong and public support for climate action keeps growing. This is why carbon reduction efforts – and climate change mitigation generally – remain such a strong focus, and rightly so.

Whatever happens, the world simply has to achieve 'Net Zero' emissions (a situation in which the amount of greenhouse gases being pumped into the atmosphere is equal to the amount of those same gases that are removed from the atmosphere by trees and plants, and various 'carbon-sucking' technologies). The mitigation effort is vital, and there is huge value in avoiding every tenth of a degree of additional global warming. Topping out at 2.1°C is better than topping out at 2.2°C or 2.3°C. 'Net zero' – better still 'Real Zero' (a situation where fossil fuel use declines to zero) – needs to be hit. The world needs to achieve 'Zero' fast – not necessarily as fast as *humanly* possible (because some pathways are far less just than others), but none the less fast – as fast as *humanely* possible.

However, even if greenhouse gas emissions are pushed into a hasty descent between now and 2050, with warming kept below a 2°C rise, climate change will still get worse before it (hopefully) gets better. In this *best-case* scenario, the journey from today's 1.2°C[8] of warming up to a *just below* 2°C peak and then back down again

## INTRODUCTION

will take decades - i.e., most, if not all, of the rest of your life. The IPCC's (the UN's Intergovernmental Panel on Climate Change) '1.5°C Special Report' revealed just how dangerous warming in this range will be: it will have devastating effects for billions of people; the next few decades will be tumultuous[9].

In worse scenarios, global temperatures will keep rising, potentially right up to 3°C above the pre-industrial average by the 2090s. Zeke Hausfather and Glen Peters, two of the world's leading climate scientists, assign a probability of *'likely, given current policies'* to the 3°C scenario[10]. Under that dreadful case, the risk of passing several key climate 'tipping points' grows significantly[11]. We risk hitting temperatures that would trigger irreparable change to forests, mountains, ice caps, oceans, monsoon patterns, permafrost and coral reefs. Scientists can only speculate on exactly what passing these various tipping points will lead to and there are disagreements about how much warming is needed to reach them[12]. However, given the strong possibility of domino or cascading effects[13] - where, for example, accelerated polar ice melt changes ocean circulation patterns, which, in turn, disrupts the timing of the monsoon season in the tropics - it is safe to assume that passing 3°C will cause highly disruptive change and great suffering.

But, worst-cases aside, the 'best-case' scenario is bad enough - a fact that needs to be stressed repeatedly. With just the 1.2°C of warming experienced so far, climate change is already disturbing and destroying millions of lives. Over the coming decades, as the climate crisis intensifies, global average temperatures will still *always* be higher than they are today. Equally, if things *do* start to get better, average temperatures – this century – will still *always* be higher than they are today. We can expect ice melt, heatwaves, superstorms, fires, and droughts to continue and to be at least as bad as they are today.

It is therefore safe to say that over the first part of this hoped for *'getting worse, then getting better'* period (the next few decades), climate change will be relentless - even under the 'best-case scenario'. It will cause at least as much damage as it does today, and it won't ease up. Billions of plants and animals will be impacted, as will the natural and built environments of a multitude of places we call home. It is not an over-exaggeration to say that hundreds of thousands of lives, and thousands of species, could be lost to climate and ecological breakdown between now and 2050. Ancient structures, bridges, buildings and roads of huge architectural, archaeological, historical, and cultural value will also likely be damaged or destroyed. Billions of people will be impacted, year after year, decade after decade. This is the argument for pursuing mitigation *and* adaptation - it is only by doing both, in tandem, that these losses can be minimised.

---

Predicting the future is a fraught business, but this book needs the grounding of two key working assumptions about where the science and politics of climate change are taking us.

**INTRODUCTION**

## Two working assumptions

Climate change is going to get progressively more dangerous. Without dramatic shifts in policy and action at a global scale, temperatures will almost certainly go beyond the 1.5°C threshold, and quite possibly beyond 2°C too. Either way, **billions of people, animals and plants will be adversely affected by climate change** - that is this book's first working assumption.

In the face of these dangerous levels of climate change, we – as a species – won't just stoically accept our fate. We will strive to prevent further warming and **we *will* adapt to the changes we're experiencing** - that is this book's second and most important working assumption.

This book is not an attempt to justify these assumptions, there is no shortage of literature on why climate change is happening and how bad (or otherwise) it could get. It is also not an argument against efforts to mitigate climate change, or that adaptation is a 'solution', or an alternative option - it isn't; it is an ally to mitigation, its strongest one.

What I am assuming is that climate change *is* going to worsen and that a growing number of humans (and other species) *are* going to adapt to its impacts. This book is interested in these adaptations, it explores the forms they will take and the knock-on effects of the choices being made.

There are questions around adaptation that cannot be ignored because adaptations have consequences, and they aren't all good (or great).

## This is a justice issue

Much has been said and written about climate change and its intersections with justice. But at the same time, there are still far too few people discussing it in these terms. Of all the forms that climate justice (and injustice) takes, three stand out.

Firstly, there is the question of who is suffering and who will suffer as the crisis deepens. Climate change disproportionally impacts on the poorest and those who have done the least to cause it, there is no justice in that. This is true in the global South, but also in the North. And, as inequality deepens, the situation looks set to worsen. In this way, climate injustice is intersecting with racial injustice: we know how disproportionate an impact climate change is having on people of colour; that can't be contested[14]. David Lammy MP is right to say that the climate crisis is colonialism's natural conclusion[15]. Given this, it is becoming very hard not to see racism embedded in the continued shortcomings of those in power who have promised climate action but delivered so little of it. And, as Elizabeth Yeampierre and others[16] argue, the environmental movement isn't entirely innocent in this regard: climate change, let alone climate justice, have not always been at the top of the environmentalist's agenda; indeed, for many it still isn't.

Secondly, there are the injustices that could result from a transition to an 'adapted' and 'net zero' economy that is poorly planned or badly executed. Most pressingly, alternative employment opportunities that are meaningful and desirable need to be found for those whose current jobs are no longer viable. If they aren't, we can expect more 'gilet jaunes' style protests of the sort that have shaken France. There could also, for example, be strong resistance from farmers and rural communities who fear that their entire way of life is under threat from 're-wilders' and others who want to limit the extent of livestock agriculture. More broadly, as the transition processes develop, there are hundreds of cultures, traditions, landscapes and habitats that need to be carefully considered, conserved, protected or enhanced. A *just transition* is vital.

Thirdly, there is the huge injustice of inflicting the impacts of climate change on those who have done the least to cause them - and then compounding this by doing next to nothing to prevent further damage, enable adaptation, or provide reparations. The prevailing and widening 'Adaptation Gap' (the difference between how much finance is needed for adaptation efforts and the amount that is forthcoming) quantifies how deep this injustice is[17].

---

Survivors of climate injustices need compensation for losses and damage they suffer as climate change hits, and funds and financing that will enable them to adapt and transition. Survivors also – simply – need the abuse to stop; continuing to knowingly hit people with climate change is akin to continuing to blow cigarette smoke in the face of someone you know has lung cancer.

To prevent and undo these injustices we need leaders who are truly committed to the climate justice agenda. Stated commitments to a cause are not enough; they need to be backed up in law and the redistribution of wealth and power. Money is needed to re-build, retreat, adapt, or to put disaster risk reduction strategies in place. Any hope of securing a transfer of power and

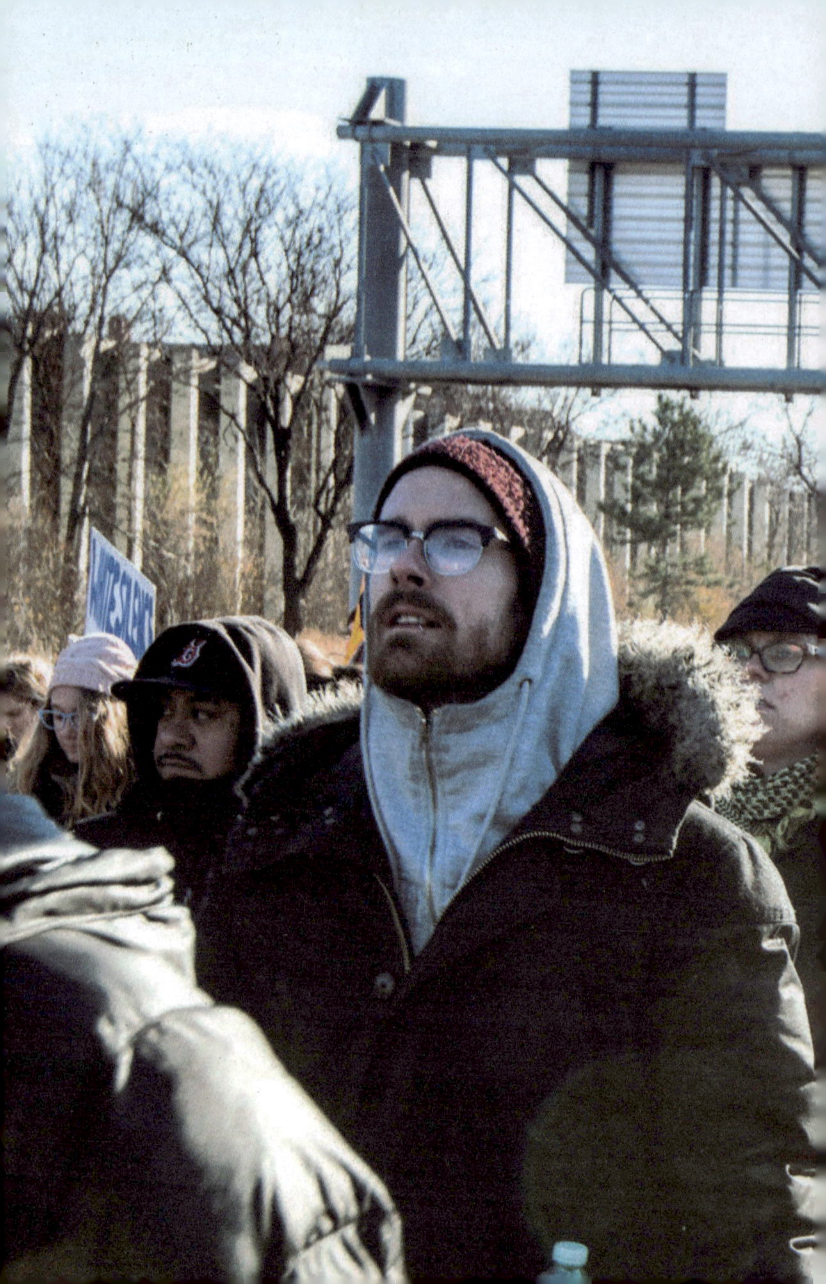

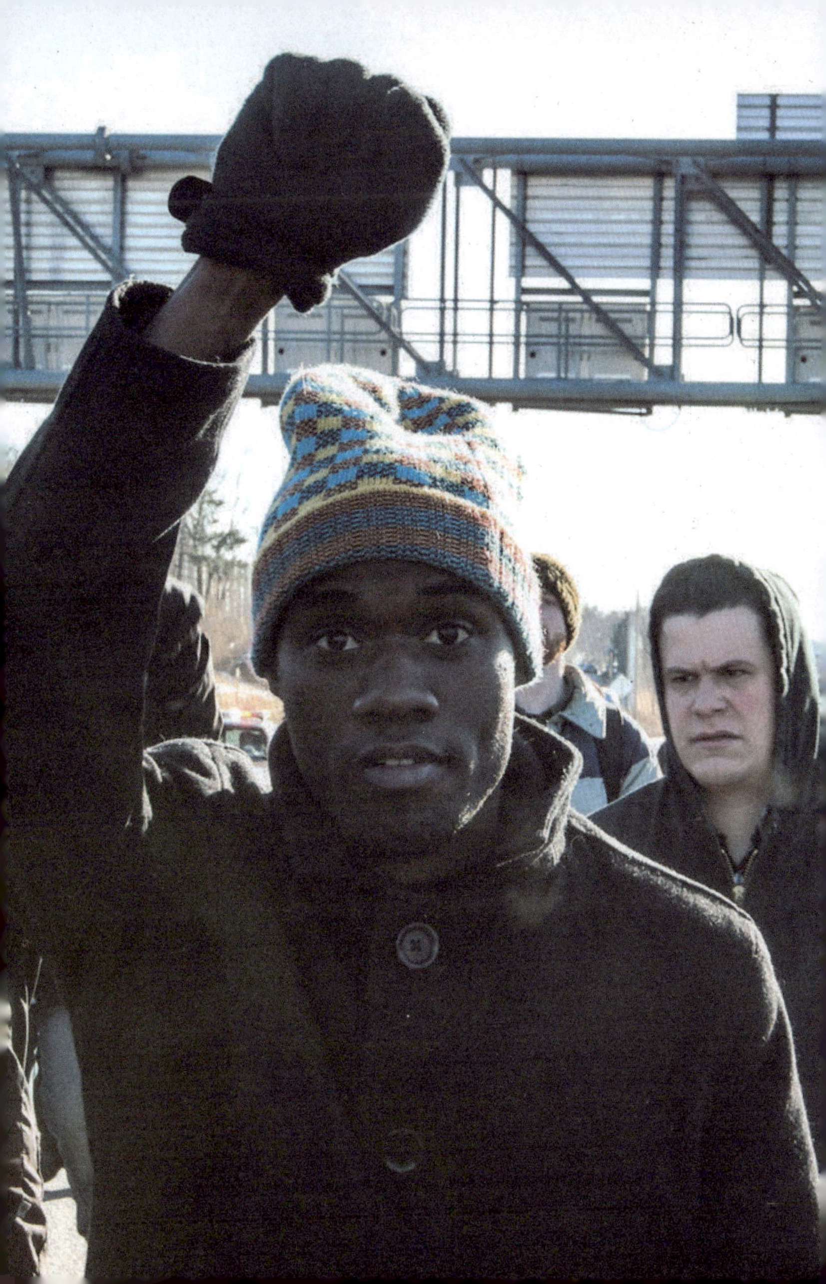

true restorative justice is tied up in the funding arrangements made. Grants must be free from the neo-colonialism that strict and onerous terms and conditions can impose. Loans must be made on terms that are favourable to the loanee and free from political motivations and private-sector profiteering.

However, countries in need of funds do not always need to seek them from external sources. Cancellation of historical and unfair debts would save countries millions of dollars every year. This money could be put to use on climate change mitigation and adaptation projects. Similarly, an overhaul of predatory international trade arrangements would enable nation states in the global South to grow their economies. Currently, as Jason Hickel has shown, the amount of money that flows into the global South – in the form of donations and foreign aid – is dwarfed by the wealth that is extracted by global North governments and multinationals in the form of the profits big businesses siphon off into offshore tax havens[18]. Root and branch reform of international trade arrangements and laws could tackle these profound injustices and allow global South countries to generate the money they need to take climate action on their own terms.

Beyond money – because money can't solve every problem – we all also need to be free to find refuge and a new life in a country of our choosing if we want to – or are forced to – migrate because of climate change. I say 'we' because all of us are going to be impacted by climate change and may, one day, need to relocate because of it.

Finally, we need policy instruments, laws and protections, that give us access to the physical materials and skills we need to mitigate and adapt to climate change in ways that *we* choose, based on *our* expertise.

---

Sadly, the injustices do not end there. To reiterate: broadly speaking, those who do the *most* to cause climate change feel its impacts the *least* - for they have the resources to adapt. Those who do the *least* to cause it suffer its impacts the *most* - for they are *more likely* to be in its path, and *less likely* to have the resources to cope. The rich adapt with comparative ease, while the poor struggle to adapt at all; the wealthy therefore prosper as the poor get weaker. The result – in an uncompromising free-market economic system – is increased inequality both within and between nations.

We also need to consider that adaptation is not a cost-free exercise; it has its own impacts on society and on the environment, not all of them are good. In choosing an adaptation strategy, one must also choose whether to consider its knock-on effects. When we choose not to, prioritising our own adaptation needs over the wider needs of others (including other species), we risk contributing to the heightening of inequality, injustice and environmental decline.

In the same spirit in which we call for a *just* transition to a low-carbon society, we must also call for *just* adaptation to climate change. They are two sides of the same coin.

## Emerging from the shadows

The human species' first major adaptation to climate change came around 12,000 years ago as the cold of the Younger Dryas gave way to the warmth of the Holocene. Slowly, over several generations, our ancestors transitioned from a life of nomadic hunter gathering to a more rooted life of subsistence agriculture, eventually establishing settlements on the newly fertile lands between the rivers Nile and Euphrates. This was a dramatic and transformative change in the way humans lived, worked and related to each other and the environment.

So, whilst we should be alarmed about the onset of climate and ecological breakdown, and mourn the extinction of plants, animals, insects, habitats and landscapes, we should not give up. Nor should we retreat into doom and despair. Despite the dangers it faces, the human species will adapt and endure. It won't be easy, it won't be without loss, but it will happen.

The task is to ensure that adaptation efforts are coupled with mitigation efforts and that both are part of a broader evolution to better ways of living and relating. This is not a time for building bunkers and lifeboats, reinforcing borders and shoring up the wealth of the 1%. It is a time for openness, cooperation and justice, these are the bedrocks of peace, and it will be a struggle to transition to something new without them.

---

We are once again experiencing a changing climate, but this time it is human-made and happening fast. The warmth of the Holocene is giving way to the heat of the Anthropocene and as it – violently – happens, we face many complex choices. We know that, through mitigation, we can delay the rate of change, and that through adaptation we can adjust our ways of living to minimise the disruption. But we are forever uncertain of what exactly to do.

In deciding how, when and where to adapt, it must be recognised that trade-offs will be made. Budgets are not bottomless, political incentives come and go, and the choice of one path can close off another. However, once a choice has been made to address climate change (not in today's half-hearted way, but properly, with tenacity and full commitment), there is then a need to balance how much priority is given to all the different forms of adaptation and mitigation that are possible – and, in some cases, whether the last £10 million of a country's annual climate action budget goes to a mitigation project or an adaptation project. These are very difficult choices, but they are the ones that lie ahead. And in the end we may just have to do as Donna J. Haraway advises: *'stay with the trouble'*[19] and muddle our way through.

---

Adaptation has long been in the shadow of mitigation, but it is slowly emerging. There is an increasing recognition of the need for not only more, but *better* adaptation. By shining more light on its intricacies and peculiarities, we can truly understand its drivers, successes, and failings. Doing this will help us make the best possible adaptation choices and ensure that the legacy of today's early adapters is not only more and better adaptation, but adaptation that *advances* social justice, *enhances* the environment, and *topples* the status quo.

This book explores adaptations that are already underway, it brings the early adapters out of the shadows. It tells the stories of people, animals, environments, institutions, and communities who have begun to adapt. It covers the good, the bad and the ugly of twenty first century adaptation. Some efforts are absurdly silly, some are brilliant, others are deeply troubling. They are all illuminating.

If you lacked the case study examples and language to campaign for adaptations that are just and transformative, and against adaptations that only serve the interests of the few, this book provides them. Telling these stories will help us lobby for more, and smarter, adaptation. If you are new, or relatively new, to climate change adaptation, this is – hopefully – a way into a topic that is often fascinating, sometimes infuriating, but definitely urgent.

---

## INTRODUCTION

*Great Adaptations* is divided into four main parts. Part one, *'Silence'*, recounts the story of managed retreat on Staten Island, New York, to explore the agnosticism that sometimes exists around climate change and climate change adaptation.

Part two, *'Adaptation'*, sets out the case for talking about adaptation and then... talks about it. There are chapters on the adaptations being made by individuals, communities, businesses, institutions, and governments - as well as the adaptations being made by wild and feral animals.

Part three, *'Transformation'*, goes deeper. It explores the notion of Deep Adaptation, the emergence of Transformative Adaptation, and what might happen if climate change gets *really* bad.

Finally, part four, *'Stories'*, looks at the reassuring stories being told about climate change, specifically in the UK, and how vital it is for adaptation that those stories are told accurately.

# PART 1
Silence

> *Mother Nature may be forgiving this year, or next year, but eventually she's going to come around and whack you. You've got to be prepared.*

Geraldo Rivera

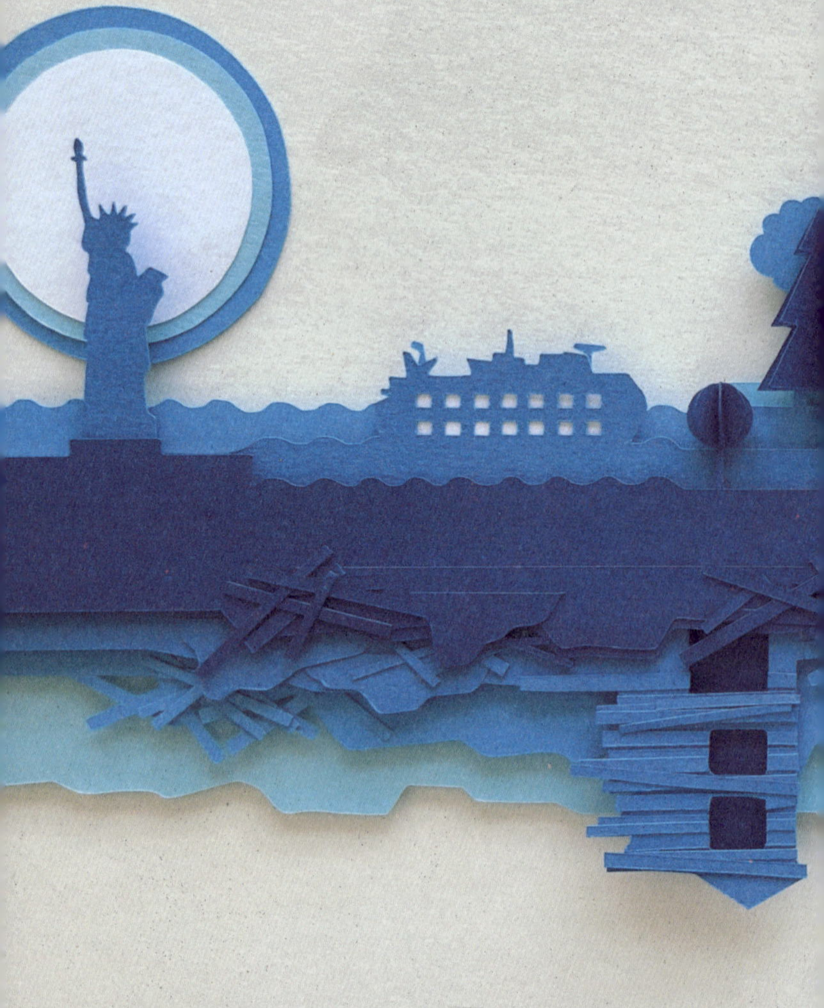

Mother Nature Wants Her Land Back

## 1. The storm before the calm

Devastating hurricanes and superstorms have become an almost annual occurrence in North America. Sandy, Katrina, Irma, Harvey, Maria and Zeta have reshaped the American consciousness. For many millions of people, the threat of devastating extreme weather is now very much a present one. Superstorm hurricanes are no longer the 'once in a hundred years' events they used to be.

On October 22nd, 2012, a tropical storm formed over the Caribbean Sea. Over the next few days, it began to travel northeast, tracking a few hundred miles off the east coast of North America. It grew, becoming a hurricane - Hurricane Sandy. As it got closer to making landfall, Sandy coalesced with several other storms. It was now not just a simple hurricane; it had transformed into a mutant Superstorm. Eventually, on October 28th, it careered towards the shore and towered over the eastern seaboard, hitting Atlantic City and New Jersey before migrating north to pummel Staten Island and the rest of New York City. The impact was disastrous; it was a wrecking ball of wind and water. One resident described it as *'like Niagara Falls becoming horizontal.'*[20]

This chapter focuses on Staten Island and its recovery from Superstorm Sandy. What happened there is a fascinating insight into how climate change adaptations play out. It is also something of an allegory of a wider story about adaptation and our willingness to accept the need for it.

---

Staten Island lies south-west of Manhattan and Brooklyn. It is New York's least populated, safest, most suburban, most white and most Republican borough. Its shoreline properties are some of the most desirable in the New York area. Open spaces, large gardens, beaches, and sea views, all within a commutable distance of downtown, make it a very attractive place to live. It is also, however, extremely vulnerable to climate change.

With a violence that is hard to imagine, Sandy destroyed hundreds of thousands of homes. In total, across eastern USA, an estimated 650,000 buildings were destroyed or damaged, millions more properties lost power and water supplies. Sandy killed 72 people directly and a further 87 died as a result of hypothermia, carbon monoxide poisoning and accidents that occurred during the clean-up. Staten Island was not spared, the wealth and political clout of its residents were no match for the storm. But, unlike many other places, post-Sandy, Staten Island had options. The option they eventually chose isn't hugely remarkable, what is of interest – as local academic Dr Liz Koslov uncovered – is the way they justified the decision they made.

Koslov spent a lot of time on Staten Island investigating its response to Sandy. She attended community meetings and interviewed various stakeholders. Her interest was less about what Staten Island did post-Sandy, but why certain choices were made. Her analysis is fascinating[21]. In simple terms, two choices faced those whose homes and properties had been destroyed by Sandy: Rebuild *or* Retreat. In the months after the initial clean up, this was the debate. But soon, rumours of State funded 'buyouts' to enable a 'managed retreat' started to emerge. For property owners, the buyout plan would see them receive the pre-storm value of their property from the State. The State would also support them to move inland to safer ground. A consensus started to develop around the idea, which is when things started to get curious.

In November 2012, very soon after Sandy hit, Orrin H. Pilkey, a Professor Emeritus of Earth and Ocean Sciences, made a passionate plea for retreat in the New York Times. He cited how the warming oceans would make Sandy-like storms more likely, frequent, and deadly. He lamented the *let's come back stronger and better* attitude, called the urge to rebuild understandable but 'madness', and signed off by urging New York and New Jersey officials to consult with the oceanographers, coastal ecologists and others who understand rising sea levels before making any decisions. His advice was clear: *'We need more resilient development, to be sure. But we also need to begin to retreat from the ocean's edge.'*[22]

Pilkey was making the climate change case for an adaptation strategy of retreat, he had allies but, contrary to expectations, the climate change case became increasingly less prominent in the debate as calm replaced the storm. Instead, Staten Islander's and their political representatives started making other arguments for managed retreat; in fact, there seemed to be a hunt on to find *any reason other than climate change* to make the case for buyouts and retreat.

Koslov observed how a narrative was emerging about *mother nature wanting her land back*. This argument gained traction quickly, it seemed to suit most people and politicians in a way that the climate change argument didn't. So, whilst for many there was little doubt that climate change multiplied the risk of future hurricanes and storm surges, a code of silence developed. Climate

When we analyse an event like Sandy and inquire into how a community recovers from such a devastating event; it is important to remember one key thing: in New York City recovery *is* possible; the City has the means to respond.

This is an advantage that is missing for so many other sufferers of extreme weather events. In Mozambique, in early 2019, two catastrophic tropical cyclones, Idai and then Kenneth, hit during the same hurricane season. Two years later, nearly 100,000 people were still waiting to be resettled when another cyclone hit, Eloise, which led to more displacement, disruption and loss of life[23].

The capacity of the US Government means that when disaster strikes, recovery can happen swiftly; not that it always does of course. The horrors of post-Katrina New Orleans teach us how, despite its vast wealth, the US Government is not always as swift, or as humane as it should be. It can spectacularly fail thousands of people during and after catastrophic events. When Katrina hit New Orleans, the US Government treated its citizens shamefully. Inequality exists between countries, but it also exists within them - Staten Islanders had the means to recover; many others don't.

G
"Moth
Her
Bu
And G

v. Cuomo
r Nature Wants
AND BACK."
US OUT
ve it Back

change was the factor that was left unspoken; to raise it was to endanger the chances that the 'buyout' plan would be enacted.

In 2014, law professor Katrina Fischer Kuh coined a new phrase to explain this silence. She was describing how individuals were beginning to adapt to climate change but doing so without knowing – or sometimes acknowledging – that it was climate change they were adapting to. She called this *'agnostic adaptation'*[24]; it can be accidental or deliberate. Koslov is convinced that what she witnessed on Staten Island was agnostic adaptation of the *deliberately not acknowledging* variety.

---

Looking for a way to respond to Sandy, New York State's Democratic Governor Andrew Cuomo latched onto the *mother nature wanting her land back* argument. The Government clearly needed to do something and, more pressingly, he required a politically palatable reason for doing it. After tentatively citing climate change in speeches and interviews, Cuomo soon pivoted and began talking a lot less about climate and a lot more about *giving mother nature her land back*.

The 'mother nature' argument was built on the foundations of long-standing opposition to the development of housing on Staten Island's low-lying coastal wetlands. Community groups had been arguing the case for less development and better flood protection since an unnamed *'Nor'easter'* had caused huge damage in 1992. Superstorm Sandy vindicated their warnings.

The *'give her land back'* campaign that rose up post-Sandy seemed to have a quasi-religious feel. Mother nature was framed as a higher power made angry by the recklessness of humans 'taking'

*Her* (wet)lands. Sandy was *Her* messenger, *Her* way of telling Staten Islanders that she wanted *Her* wetlands back - Staten Islander's became determined to *'give it back'*.

Koslov's paper is a captivating exploration of the politics of silence. She identified many nuanced reasons why people from all different backgrounds decided not to talk about climate change and opt instead for mother nature:-

- Although many were in favour of managed retreat, there wasn't total agreement across the community that it was the best option; some wanted to rebuild. Climate change was seen as a polarising issue. So, in order to minimise conflict and sustain 'retreat' group harmony, it felt wise to avoid it as a topic. Discord within the group would threaten the united front seen as vital for securing a favourable result (managed retreat) in negotiations with the government.

- This part of Staten Island is politically conservative and neo-liberal; government intervention is generally viewed with suspicion, something to be discouraged. But every rule has an exception, 'buyouts' looked attractive, so government intervention was now, suddenly, fine. A legitimate reason for asking for it was therefore needed. By framing greedy developers and the politicians who supported them as rogue actors, in an otherwise functioning system, they found their reason. The flood risk was presented as the fault of rogue actors, it was *they* who had taken the land from mother nature and *they* who had allowed the flood plain to be covered with impermeable concrete surfaces. Intervention to right previous wrongs was therefore warranted in this special case.

- Linked to this, there was a feeling that climate change might give politicians and other powerful actors a get out of jail free card. If the focus was on climate change it would enable powerful actors to avoid responsibility for creating the problem and therefore for funding the solution.

- Governor Cuomo seemed willing to go along with this scapegoating of rogue actors from the past, but for a different reason. He knew that if climate change was positioned as the pivotal argument for managed retreat, he would find it very hard to narrow the geographical scope of his responsibilities. It therefore also suited him to focus on the *'mother nature wants her land back'* narrative; the amount of land she wanted back was considerably smaller than the amount of land threatened by future climate change powered superstorms.

- This narrowing of the geographical scope also helped Staten Islanders who were worried that Sandy survivors from Queens and other affected areas would compete with them for the limited pot of funding available for buyouts. If localised bad planning was the reason for managed retreat, rather than climate change, claims for buyouts from Queens and elsewhere would have little validity.

Staten Island wasn't adapting to climate change, it was righting the wrongs of historical flood management plans!

- In the wake of Sandy, environmental justice campaigners were calling for systemic change to help marginalised communities in New York (and elsewhere) secure the financial and political support needed to deal with the everyday impacts of climate and ecological breakdown they are already suffering. Koslov argues that the wealthy, conservative, neo-liberal residents of Staten Island's east shore mostly oppose action on climate change of the sort proposed by environmental justice campaigns. So, for them, finding an argument for managed retreat that didn't mention climate meant that they could avoid alignments with pressure groups whose politics they were opposed to. Removing climate change from the argument for managed retreat starved environmental campaigners of oxygen; a bonus outcome that suited the broader political objectives of Staten Island's more politically conservative residents.

- It was not just wealthy Staten Island residents who saw benefits to remaining silent on climate change; low earners seemed keen to avoid the topic too. They worried that a focus on climate change would lead to calls for carbon taxes, insurance price hikes and more fundamental systemic change. As Koslov puts it: *'those who, while reaping diminishing benefits from the [economic] system as it stood, also felt that they could ill afford its dismantling.'*

Some of these justifications for agnostic adaptation are easier to sympathise with than others. When there is very little money left over at the end of the month, the threat of an additional tax can feel existential. In societies with high levels of income inequality, it is unsurprising to hear those near or below the poverty line expressing serious concern about government imposed increases to their costs of living. The other reasons for remaining silent that Koslov teased out sit less comfortably. The behaviours she describes range from nose holding political pragmatism, to outright selfishness.

Managed retreat is now well underway on Staten Island. The ocean facing neighbourhood of Oakwood beach has been largely abandoned. Its former citizens have used their buyout money to purchase homes inland. Where there were once busy houses, roads, gardens and swimming pools there are now trees, flowers and wildlife. Oakwood beach is not quite a ghost town, but mother nature *has* got her land back and Staten Island is now a little bit more resilient to climate change. A lot can be learned from this case study, but it can also be used as a mirror to reflect on the stories being told (and not told) about climate change.

## 2. Code of silence?

On post-Sandy Staten Island it became taboo to talk about climate change; people declined to publicly acknowledge its importance, or the role it was playing in the decision-making process.

For climate change deniers, who had spent their lives trying to ignore, dismiss, refute and play down the level of risk attached to the continued release of greenhouse gases, it was unthinkable that they would admit that their retreat from the shoreline was based on their concerns about climate change. Such an admission would have consequences, and they knew it. It would humble and embarrass them, denting something core to their identity. It would also embolden their political rivals.

Climate change 'believers' resolved that it was more pragmatic to let the 'deniers' tell themselves (and the world) that the case for managed retreat had nothing to do with climate change. Not talking about climate change became the politically expedient thing to do. The case had to be based on something else - *anything else*.

As understandable a strategy as it was, the climate change 'believers' on Staten Island who went along with the deliberate silence on climate change were, in effect aiding denialism, while also implicating themselves in another negative consequence of agnosticism. Staten Island's Oakwood Beach is just one of many communities along the US coastline that is threatened by rising sea levels, hurricanes and superstorms. Homeowners in other coastal communities will be looking at Oakwood Beach with sympathy of course, but also some envy. They will have seen how buyouts cushioned the devastating blow dealt by Sandy, while they experienced a much harder landing.

However, because climate change was missing or downplayed by those making the case for retreat, campaigns that cite Staten Island as a case study to be repeated are far less powerful than they should be. This is the downside to the story told by Staten Islanders; it denies campaigners the option of pointing to what happened there as being a replicable example of climate change driven managed retreat. They are not able to argue that Staten Island is a precedent. If they were able to, they could ask the US Government: why, if Oakwood Beach was deemed at risk of climate change catastrophe, isn't their community also worthy of the support needed to retreat? But they can't ask those questions because – officially – managed retreat on Staten Island isn't driven by climate change. Staten Islanders were simply *'giving back to mother nature'*.

This is what makes *deliberate* agnostic adaptation so distasteful. Refusing to acknowledge climate change as a factor in decision making, refusing to even talk about it – just to protect one's own self-interest (whether you are a 'believer' or 'denier') – has repercussions for others. Managed retreat on Staten Island unequivocally is an adaptation to climate change, but it was an adaptation story that wouldn't be told.

---

On behalf of all those people, all over the world, who are crying out for support for managed retreat – as a way to adapt to climate change – we should be angry with the agnostic adapters of Staten Island. But we, who are part of the climate movement, should also look deep into the mirror and ask ourselves if we are just as agnostic about adaptation, as Staten Islanders were about climate change?

- Are we acknowledging the climate change adaptations that are happening all around the world?
- Are we reporting on them?
- Are we asking authorities to invest deeper and faster into projects that will enable the most vulnerable to adapt?
- Are we pointing out the precedents?
- Are we ensuring that those who are making adaptations aren't doing it in selfish ways, ignoring the knock-on effects of their projects on neighbouring communities or future generations?
- Or, are we staying quiet?

# PART 2
# Adaptation

> *We are just a group of ordinary people from the Warwick District. We are a diverse group who were randomly selected to join this Climate Change Inquiry. We have different viewpoints but have reached a shared understanding. We now recognise we are in a climate emergency. We must act now all together and with urgency. Through our conversations with experts and with each other we have come to recognise the importance of immediate action. We are not experts, and we recognise that it may not be 100% possible to implement all our recommendations immediately, however we ask all local organisations and institutions including Warwick District Council to take our recommendations as a mandate to be as ambitious as you can, within the areas we have highlighted, in responding to the emergency we face.*

*This is everybody's problem, everybody has a role to take action. We are just a small group of people but through our coming together we have become a group bursting with ideas and enthusiasm. We believe that to respond to our emergency the District Council and other organisations must harness the energy and enthusiasm of our people and our communities. We must all make a change for climate change. Your district needs you.*

*We believe the solutions are out there already we just need to make it happen. If we can't find the examples, then we must be those examples.*

*The District Council and others must look for every opportunity to influence all who have a role in our district. We must make this happen and the frameworks must be in place to make sure there is accountability (named people) and monitoring (performance indicators) so that everybody can check progress.*

*THIS IS AN EMERGENCY.* **"**

Joint statement of the members of the District of Warwick people's inquiry on climate change, 2021[25].

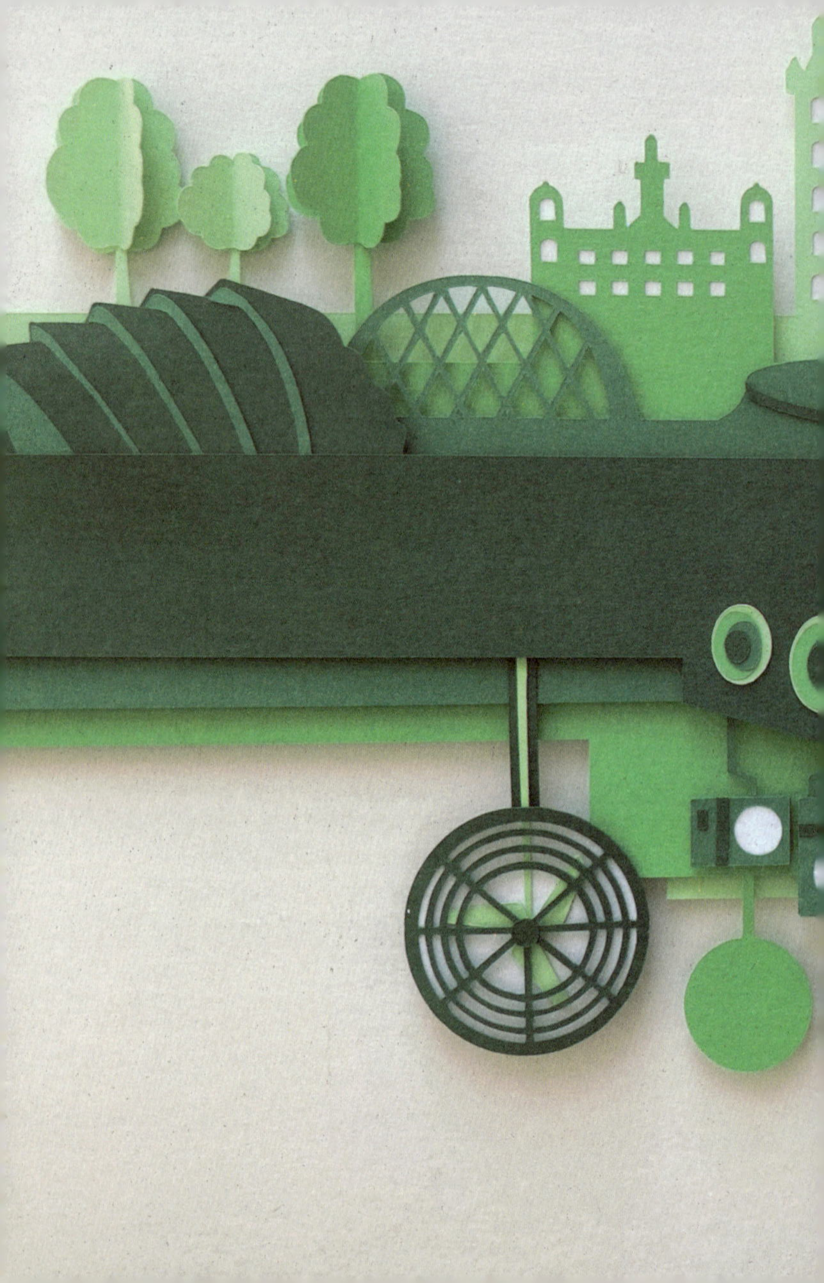

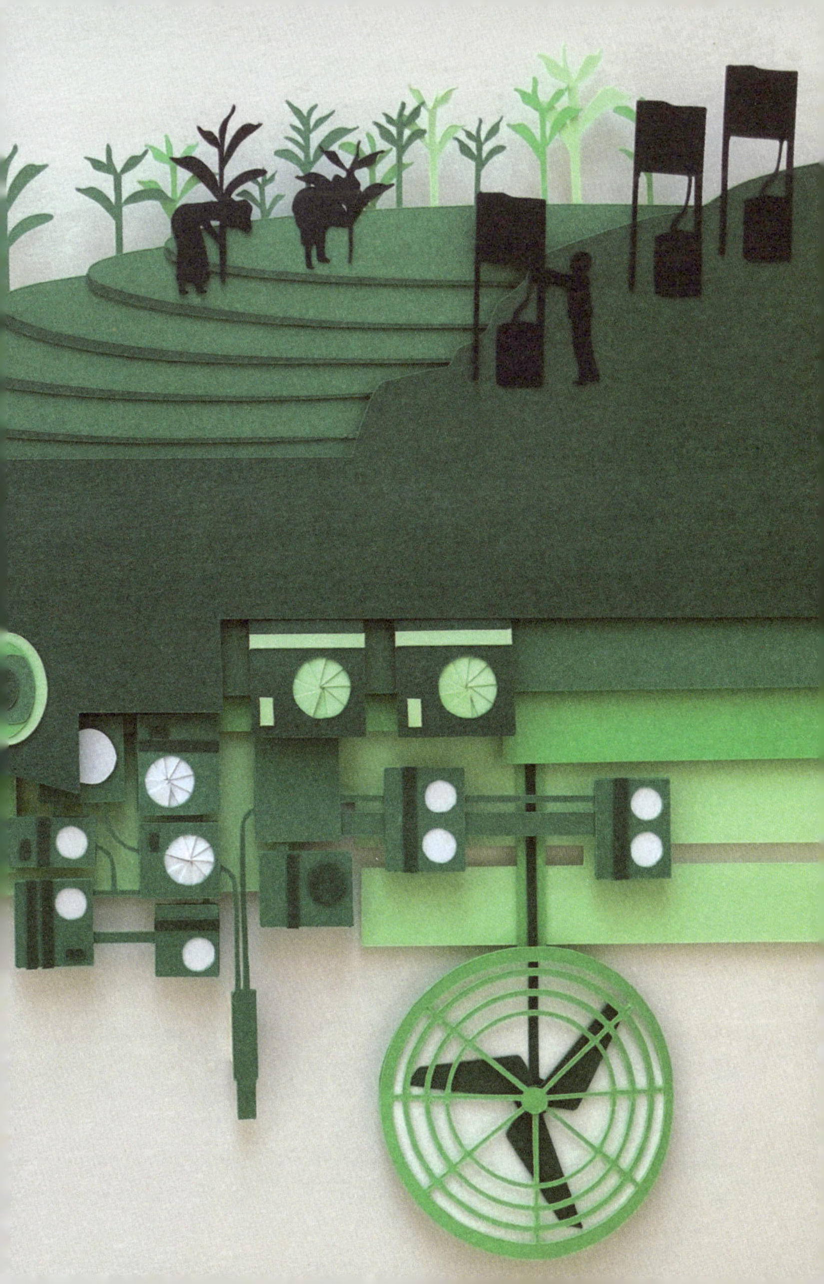

# 3. Five reasons to talk about adaptation

If you are sceptical about the wisdom of talking adaptation, maybe because mitigation is more your thing (or because you want that Staten Island buyout money for yourself), it pays to remember that the adaptation story is not, and does not have to be, in opposition to the mitigation story. Mitigation makes adaptation easier, and adaptation makes mitigation easier; they are partners. The urgent need to lower emissions, preserve forests and develop new models of democracy and economics is as much a part of the adaptation story as it is the mitigation story.

This book makes the case for adaptation as an essential comrade to mitigation and other key social and ecological causes. Regardless of what is thought of it, adaptation is going to happen - it is *already* happening. The pressing need now is to enable it to happen in ways that progress, rather than hinder, our attempts to mitigate and advance other causes. Talking about adaptation, and it's evil twin *mal*adaptation, is the essential first step in achieving this. This is *Great Adaptations'* call to action. We can't keep silent on adaptation, it cannot be allowed to go on happening in the shadows.

---

If the climate movement continues to be agnostic about adaptation, if it fails to tell adaptation stories, at least four things could happen:-

1. Adaptation will struggle to climb up the agenda of the world's most powerful decision makers. The adaptation

needs of those already suffering the direct impacts of climate change will therefore not be heard. Their efforts to lobby for the resources needed to adapt to a crisis – that most of them played no part in creating – will then continue to be ignored and marginalised. They will go on suffering while the powerful concern themselves with never ending technocracy, and their own adaptations.

2. If adaptation remains low or not even on the agenda; it will either not happen at all, with many perishing or being forced into extreme evasive action; or, as survival instincts kick in, it will happen in ad-hoc, insufficient, unplanned and often *mal*adaptative ways.

3. With adaptation low on the agenda, those with the means to adapt will still adapt, but won't necessarily be mindful in how they do it. With little scrutiny and limited accountability, the temptation (and opportunity) to adapt in self-interested ways will be strong. Agnostic adaptation of the kind observable on Staten Island is likely to become more prevalent, and more examples of the world's natural resources being used with impunity to power air conditioning units, build flood defences, and protect consumer capitalism by any means, will become apparent.

4. Those with the means to adapt and compassion to do so in mindful and *just* ways might also end up *mal*adapting too, by mistake. They will adopt strategies that are naïve, or prone to producing unintended consequences. Why? Because they haven't been exposed to information, training, best practice examples and the economies of scale that come from the

scaling up of innovative adaptation methods.

There is another danger too. If the climate movement doesn't talk about adaptation, other movements will, and this might create some unjust and undesirable outcomes:-

5. There are several adaptation narratives. For example, some people frame adaptation as an incremental process, while others frame it as a transformative one. Some think of it as an exercise in accommodating climate change, others as an exercise in resisting it. Some frame the responsibility for adaptation at the individual or community level, others see it as something for national or multinational bodies to deal with. As yet, none of these framings have become fully stuck in the collective conscience, but it is something to be vigilant about. The narratives that do end up dominating will have a powerful effect on the forms of adaptation that emerge over the coming decades. This is because the stories told about climate change adaptation, shape not only how it is understood, but also how it is done (which can be *to* and *for* people; or *with* and *by* them); stories are powerful things.

The Glacier Trust has its own narrative when it comes to adaptation: we see it as a citizen-led, mindful process that transforms the relationships humans have with each other and the natural world. We believe in *just* adaptations, and *just* transitions to new social and economic models that are fairer and more ecological. These principles are, we hope, reflected in the adaptation projects we enable in Nepal.

I will revisit the topic of how adaptation is framed in the final chapter of this book. It has huge implications, but my immediate aim is to actually talk about adaptation. On the pages that follow, you will find stories of the early adapters to climate change. I cover adaptations that are instinctive or accidental, as well as adaptations that are carefully planned and designed with very deliberate outcomes in mind. The approaches and results won't always sit comfortably with you. But I do not wish to only tell you about 'good' adaptations. It is possible to learn as much, often more, from 'bad' and 'ugly' adaptations. You will find a smörgåsbord of approaches and narratives. It is a look at who is adapting and for whom.

My primary aim is to highlight *what* is happening, while only touching on the *why*. These deeper questions, and questions about the implications of the adaptation choices being made, are what I hope *Great Adaptations* might spark. In the language of my environmental education colleagues, these stories are stimulus materials.

# 4. Air condition everything

58% of the world's population live in urban environments[26]. Over the coming decades, this proportion is set to rise. Global population will also rise (albeit at a declining growth rate), so, over time, an ever higher percentage of a larger population will be living in towns and cities. By 2050, approximately 7 billion people will be urbanites, up from today's 4.2 billion[27].

Already the majority of human beings are experiencing the impacts of climate breakdown while living in an urban environment. How urbanites cope and adapt will be the climate story of the twenty first century. Air conditioning looms large, and this chapter delves into it, but it is not the only adaptation strategy we're seeing - which is why this chapter starts in Glasgow.

## Welcome to Glasgow

Having been given the responsibility to be the main co-host (with Italy) of the UN's twenty sixth Conference of the Parties (COP26), the UK Government eventually chose Glasgow as the venue for the most important COP since Paris 2015. To call COP26 long-awaited would be an understatement. The agreements and commitments made in Glasgow will set the agenda for climate action in the coming decades. The action needs to be radical, and it needs to be soon. In recent years, the encouraging rhetoric on both adaptation and mitigation has steadily grown, and *some* signs that *some* of the major polluting countries will deepen their commitments are visible.

Despite this gradual gathering of momentum, few expect to see a near-term flurry of sweeping reforms and real world activity. Mitigation efforts are unlikely to radically upscale – not in the 2020's at least – and there is even less chance that adaptation efforts will. The UN Framework Convention on Climate Change (UNFCCC) has never tended to inspire such radical action; their set piece gatherings (the COPs) are a grinding power struggle for almost everyone involved. This was a point made forcefully by Claire O'Neill in a candid letter to the UK Prime Minister (who had just brought her role as COP26 President to a premature end):

> *The annual UN talks are dogged by endless rows over agendas, ongoing unresolved splits over who should pay and insufficient attention and funding for adaptation and resilience.*[28]

To many outsiders, the removal of O'Neill seemed to be unfair; she was ousted by Boris Johnson who replaced her with a man, Alok Sharma MP, whose voting record on climate and ecological issues hadn't made him an obvious candidate. O'Neill's exasperations about the UN processes were, however, very insightful - a rare glimpse behind the curtain. Her frustrations about the deprioritising of adaptation and resilience are shared by many working on these issues. We are distressed, but perhaps not terribly surprised by the lack of attention adaptation receives. We cannot, after all, expect the UN and world leaders to give enough attention to adaptation and resilience when they are receiving next to no pressure on this from the mainstream climate movement. Adaptation is not something environmental organisations campaign on.

There are, however, some indications that O'Neill and others are being listened to, with adaptation showing up a little more in recent debates. Alok Sharma spoke often on the subject in early 2021, and even Boris Johnson is now on record emphasising the importance of supporting vulnerable countries *'to adapt and build resilience'*[29].

So, adaptation will be higher up the agenda than ever before at COP26. How high remains to be seen, and here at The Glacier Trust, we are not allowing ourselves to get carried away. Adaptation is likely to remain in the shadow of mitigation for a while yet, so our advocacy work is far from done.

---

If things were different, if adaptation was top of the agenda at COP26, then delegates would be clamouring to hear about 'Climate Ready Clyde', an initiative that aims to ensure the Glasgow City Region is adequately prepared for the impacts of climate breakdown. The work being done there is world leading, innovative and replicable.

At first glance, Glasgow isn't the city one would immediately think of as an obvious contender to be a global leader on adaptation planning. There are hundreds of city regions that would appear to be more vulnerable; there are several in the UK alone. Leadership on adaptation does not seem to be too closely related to the level of precarity faced. It is therefore interesting to explore how the Glasgow City Region came to prioritise it.

The first Climate Ready Clyde workshop was held in 2011. It was formally established in 2017 and is governed by representatives from fifteen, mostly public sector, bodies - all of whom have a stake in ensuring the Glasgow City Region is 'climate ready'. Climate Ready Clyde's origins can be traced back a little further, however, to Section 3 of the Climate Change (Scotland) Act 2009, which states:

> *all public bodies need to be resilient to the future climate and to plan for business continuity in relation to delivery of their functions and the services they deliver to the wider community.*[30]

In short, all public bodies in Scotland now have a legal duty to develop and implement a climate change adaptation strategy. It was on the back of the Climate Change (Scotland) Act, and the launch of the Scottish Government's Climate Change Adaptation Programme, that a new programme – Adaptation Scotland – was launched. Adaptation Scotland has been run since its inception by the wonderfully named 'Sniffer', an Edinburgh based charity whose mission is to be: *'change makers and knowledge brokers for a society with greater resilience to environmental change, in particular climate change.'*[31] Climate Ready Clyde is a product of Sniffer's work under the Adaptation Scotland banner.

Ultimately, Government policy is only as good as the people who deliver on it. Under various National Adaptation Plans around the world, public bodies have similar duties on adaptation and resilience as their Scottish counterparts. This does not mean they fulfil them. Local authorities, transport departments, health providers, emergency services and various land management, cultural and heritage bodies, struggle to give adaptation the attention it needs. In many cases it is barely getting any attention at all. This is understandable, as public bodies everywhere are stretched by austerity and an ever more complex list of overlapping and competing challenges. It is easy to see how an issue like adaptation can be side-lined. And, in a world where campaigns for more and better adaptation are almost non-existent, it certainly isn't an issue that elected representatives and public sector chief executives are feeling the heat on - not from the public, not from the media, not even from mainstream environmental NGOs.

In this context, it takes something special for meaningful action on an unfashionable – but highly important – issue like adaptation to surface. It takes the wit, intelligence, creativity and sheer tenacity of an organisation like Sniffer, but it shouldn't. Every urban area in the UK could have a climate ready action plan as good as the one being developed and implemented by Climate Ready Clyde.

---

We can take heart that Climate Ready Clyde exists - it is a blueprint that other city regions can build from. The Climate Ready Clyde website details the journey they have been on. They've now arrived at the newly launched Glasgow City Region Adaptation Strategy and Action Plan 2020-2030[32]. The plan was several years in the making and built on the development of a collaborative network of public bodies, business and other stakeholders, a careful series of consultation exercises, much research, and a comprehensive climate risk *and* opportunity assessment for the region. The assessment focussed not just on the threats of climate breakdown – storms, floods, heatwaves, droughts – but also on possibilities. James Curran, the chairperson of Climate Ready

Clyde, put it like this:

> *There is a great prize to be won. Adapting to a changing climate will help protect jobs, deliver economic prosperity, improve wellbeing, and ensure that Glasgow City Region remains a great place to live and work for generations to come.* [33]

The 'opportunity' element of the assessment was crucial. The Climate Ready Clyde team recognise the innate human desire to make progress; there is a strong emphasis on the future in all of their communications. The Adaptation Strategy and Action Plan means Glasgow City Region is ahead of most other cities, it is getting ready for whatever climate future emerges. So, whether global average temperatures are 1°C or 4°C warmer than pre-industrial times, Glasgow will be more prepared than most.

## You got *Coolth?*

Let's head south. In the summer of 2019, The Sun newspaper reported that sales of desk fans, ice creams and cooling mats for pets[34] were skyrocketing as the UK was gripped by a prolonged heatwave. Temperatures were regularly hitting the mid-thirties, with a highest recorded temperature of 38.7°C at Cambridge Botanic Garden on July 25th. It had never been hotter in the UK[35].

Like any respectable tabloid journalist would, The Sun's Helen Knapman was salivating over the exasperated tweets of a sweaty Britain. And who doesn't love a panic buying story? According to Knapman[36], Currys PC World had reported a 200% increase in sales of fans; sales were up by 120% at John Lewis (they were 'selling six fans a minute'); AO.com had seen fan sales jump by a staggering 591% compared to the week before! You cannot of course believe everything you read in the newspapers, but sales of consumables that help keep you cool undoubtedly went up by a lot No doubt, a few weeks later, instances of fly-tipped broken desk fans went up too, along with the number of ice cream wrappers littering the streets. We'll never know how many batteries found their way to a landfill site, but it takes a lot of AAA batteries to keep a legion of mini plastic fans whirring 24/7.

When it gets that hot, people react and so do companies; an opportunity to profit from people's sudden desperate need for 'hand-held fans that spray water'[37] or 'sun loungers for pets'[38] is too good to miss. In fairness, some companies were seeking to help heat-stricken customers too: in an act of apparent gallantry, Currys PC World cut the cost of some fans to half price[39].

---

Temperatures in the mid to high thirties are a rarity in the UK. Heatwaves are no longer a complete novelty, but what we experienced in July 2019 is not yet something we're used to. That's why our adaptations are often reactionary, rather than anticipatory; many of us fail to prepare, relying on instinct.

In France, intense heatwaves are a bit more common. As a result, French cities and citizens are a little more prepared. The whole of France had suffered terribly from the 2003 heatwave - 14,802 people lost their lives. That summer had a haunting effect on the French Government at National and Municipal levels. They didn't want their citizens to suffer that way again. The

French authorities and people now have adaptation and coping strategies that can be learned from, but it is fair to say they are a work in progress.

In Paris, during the same July 2019 heatwave that hit the UK, temperatures rose to over 40°C. Parisians sweltered, but did they cope? Gonzo journalism has its limitations, but Megan Clement's account of a July day spent following the Paris heatwave plan that summer is an alluring lens through which to explore how well adapted Paris is to extreme hot weather[40].

Clement is Australian but was living in Paris that year. When the heatwave hit, she decided to put the official advice of the Paris authorities to the test and write about it for the Guardian. Her day started at home in the 20$^{th}$ arrondissement. The 20$^{th}$ is in east Paris, it is edgy, hip, Edith Piaf used to live there, it has a vineyard; it is the sort of place a twenty-something freelance journalist would live - Charlie Hebdo is based there.

The first piece of advice Clement followed was straightforward enough: close the shutters. The shutters help to keep the indoor temperature down; curtains also help. Windows are usually better off closed too, but only while indoor temperatures are lower than outside. Obviously, it is not an exact science, drafts can have a cooling effect.

The French health department recommends people find low temperature indoor spaces to spend time in during a heatwave. The advice is to spend at least three hours indoors over the course of a day; morning might be the only time your city centre flat is cool enough to hang out in. However, after a hot and sweaty night, Clement wasn't going to stay indoors and neither was her dog. So,

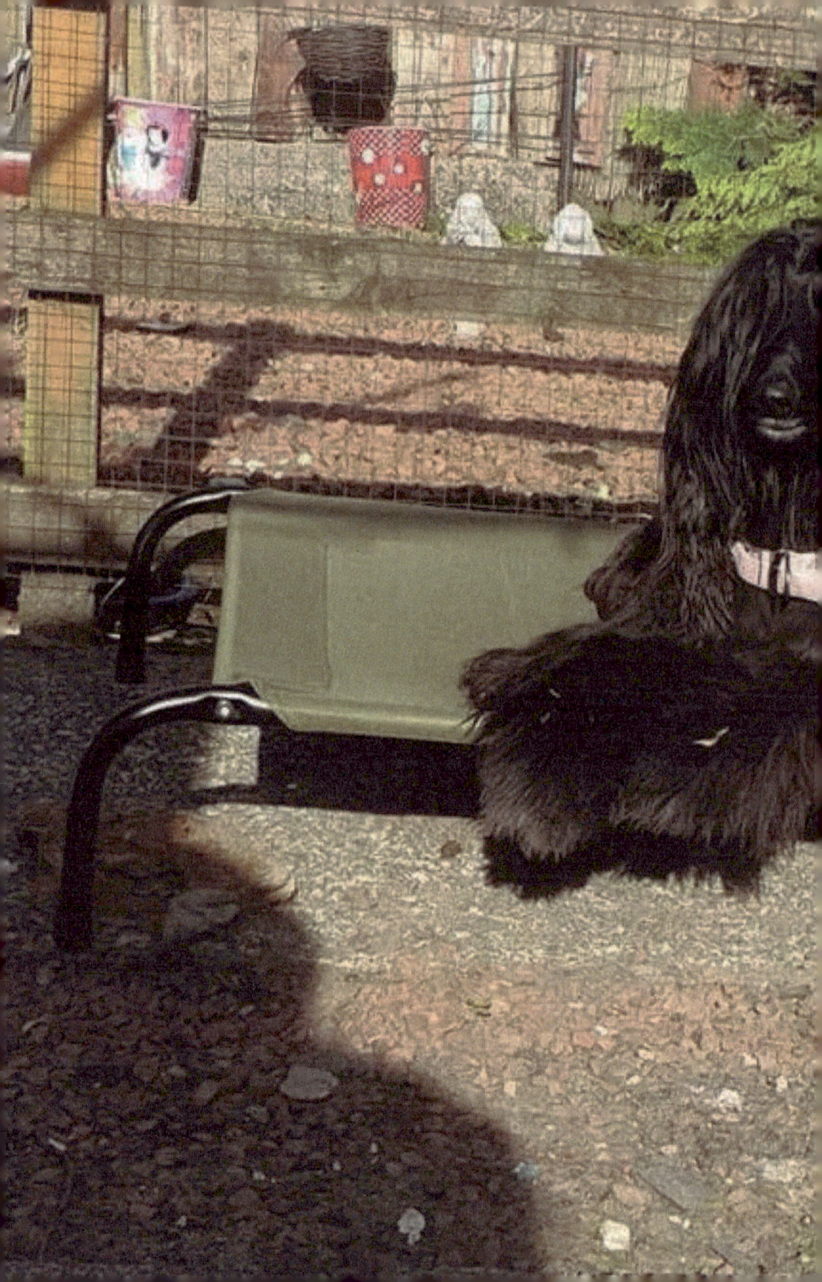

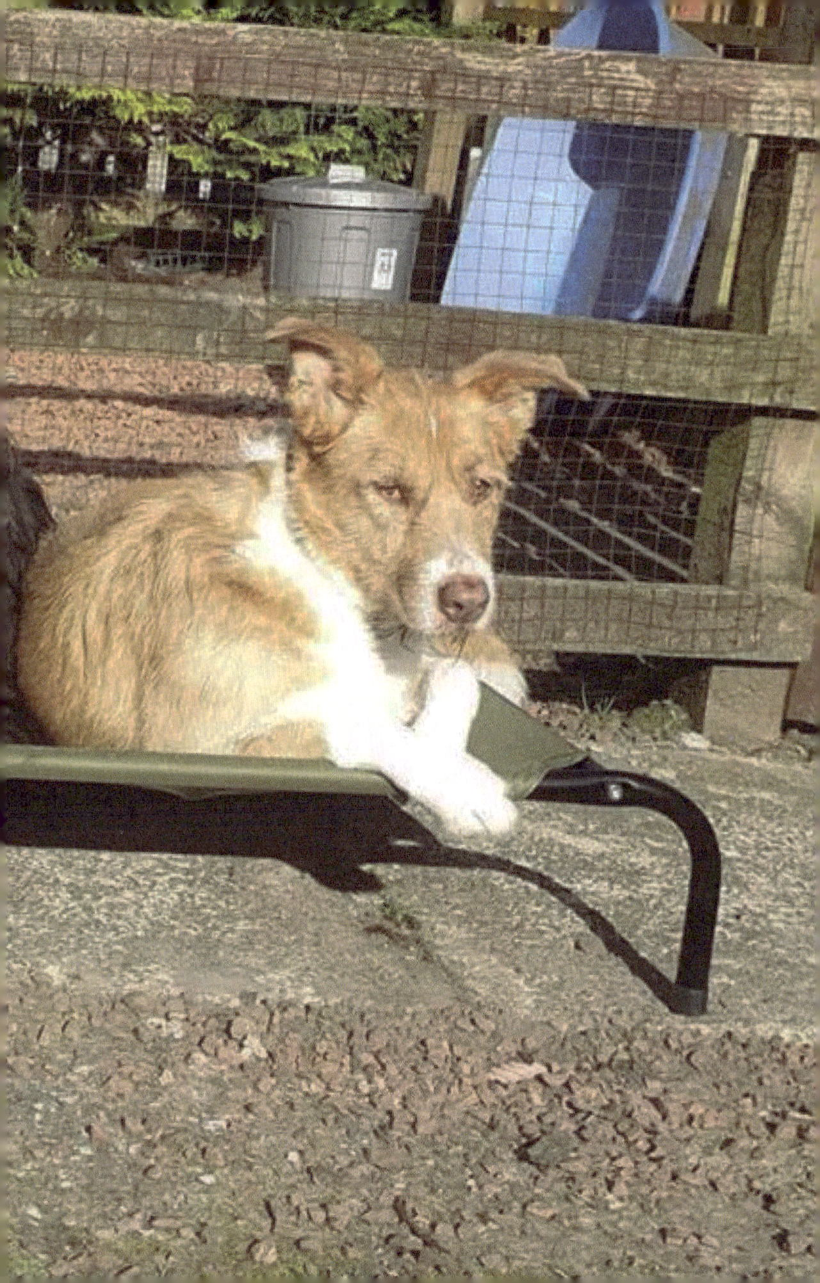

with shutters left closed, out they went for a quick stroll before the pavements warmed up too much - dogs have sensitive paws.

How do you decrease the temperature of a room? You crank up the air conditioning. What to do if you don't have an air conditioning unit? Panic buy one. What if you can't afford one? You slip into the foyer of a high-end hotel. This is what Clement and her dog did next; they were immediately confronted with an example of the inequalities of climate breakdown in micro form.

She found herself surrounded by people who had spent the night at a blissful, air conditioned, 18°C. Fair play to them, if you find yourself sleeping in a Paris hotel in the middle of one of the hottest heatwaves in history, you're probably not going to forego the coolness of the A/C. But also, 'chapeau' to Clement for resenting the woman she observed in the hotel lobby; she was wearing a scarf! Clement moved on, it's hard to keep your temperature down when your blood is boiling.

Much of the advice Clement was following came down to what can be described as a search for *coolth* (as opposed to warmth); she endured the oven-like heat of buses and metro trains to visit some of the 922 'islands of cool' mapped out on Paris' official 'Extrema' app. These include shaded parks, green spaces, libraries, churches and temporary water misters.

One of the innovations Paris has adopted originated on the west coast of America - the 'cool room' - which is where Clement headed next. Cool rooms have been set up alongside 'Chalex' - a telephone check-in service to look out for vulnerable people. Those registered with the service are monitored regularly and can be brought to air-conditioned meeting rooms in town halls if the heat where they are is getting too much. The Paris government doesn't just provide the *coolth* people need at these cool rooms, they also offer cold drinks and wet towels, as well as newspapers and magazines to help the time pass. The cool room Clement visited had a small portable air conditioning unit; it was struggling against the 42°C heat.

The rest of Clement's day saw her popping into mini-markets to press cold cans of orange juice on her face, a dip in an outdoor swimming pool, a trip back to the air-conditioned hotel for a light meal, and finally a midnight walk in one of the many parks in Paris that are now left open 24 hours a day in the summer months. Her overall assessment of the official advice?

> *I'm acutely aware that I'm healthy and have some freedom and control over my movements. Many Parisians don't: they have to move around the city according to their daily schedules, no matter the temperature. Some have no choice but to be outside all day during a heatwave. With its apps, cool spots and check-in services, the City of Paris is doing a lot to help them, but if my day taught me anything it's that reality seems to be catching up with what research has long been telling us. The effects of rapid climate change are already here - and we are not ready[40].*

## Air-conditioned pavements

If we move further south (and east) we can find a glimpse of what developed cities like Paris and possibly even Glasgow might be doing in pursuit of *coolth* in two or three decades time. Doha, the capital of Qatar, is one of the hottest cities in the world.

Its highest recorded temperature is a staggering 50.4°C, and its climate has already changed to the tune of 2°C[41]; it looks set to go on changing. Average temperatures could be 3°C - 5°C higher than the pre-industrial averages by the end of this century. This presents a huge adaptation challenge.

Doha's response is soon going to be thrust into the global spotlight. In 2022, for reasons it would take an entire other book to dissect, the men's football World Cup will be held in Qatar. It has already been moved from its usual mid-summer slot to a slightly cooler November. Putting it on, and keeping players and spectators safe, has created the highest profile *coolthing* challenge the world has ever seen. Fortunately, or unfortunately depending on your perspective, Qatar has an abundant supply of fossil fuel energy they can call upon to air-condition stadiums, hotels and, it seems, everywhere in between. So, not only will its new sports stadia have air conditioning fans under every seat, but the growing trend for installing A/C units *outside* to cool pavements, street cafés and outdoor shopping malls also looks set to spread.

If outdoor air conditioning becomes the norm in Doha, we can expect it soon to be the norm in other oil rich Middle Eastern cities too, and then why not southern Europe, the US, Australia and every other wealthy city or neighbourhood around the world? Air-conditioned pavements powered by fossil fuels - this is climate change *mal*adaptation at scale.

Again, it is easy to scorn these *mal*adaptations; but they are happening, and they need to be highlighted. Understandings of why *mal*adaptations are happening also need to be discussed and improved - it is lazy to simply play the blame game. This isn't necessarily a case of pure decadence by a handful of extremely wealthy people. Outdoor air-conditioning makes outdoor dining and socialising possible all year round in Doha, in the same way as gas fired patio heaters keep beer gardens open well into the winter months in northern Europe and North America. As climate change intensifies, the urge to maintain some normality will only increase in desert countries like Qatar, and it is easy to sympathise, the people of Doha quite rightly fear a life lived exclusively indoors. Qatari American artist, Sophia al-Maria, was expressing these concerns as far back as 2012:

> *With the coming global environmental collapse, to live completely indoors is the only way we'll be able to survive. The Gulf's a prophecy of what's to come* [42].

She might be right and, on a purely human level, the prospect of becoming a prisoner to air-conditioning is pretty bleak. On this reading, air-conditioned pavements seem a bit less irrational; the climate change driven retreat indoors has very human consequences. Very few of us like feeling trapped, and it is terrible for people and society when we are.

## Generation 22°C

Heading further south and further east again, we come to Australia, infamous for its devastating bush fires, and discover it is already adapting in similar ways to Doha. *Air condition everything* has become a mantra as urban areas adapt to increasingly long and extreme summer heatwaves. 'Cooling the Commons', a project run by a team of designers and human ecologists at Western Sydney

University, has begun to explore what climate change and, by extension, air conditioning means for life in hot cities.

Whilst the Cooling the Commons team do cover the impact of an increasingly air-conditioned world on climate change (they have produced pamphlets[43] and distribute practical advice on *'how to stay cool without contributing to climate change'*[44]), their primary interest is on its societal impacts. Their interviews and focus groups reveal to us how air conditioning is changing the way we live. They report how new houses in Sydney and across Australia are increasingly being designed around an assumption that air-conditioning is a universal standard. And it is not only housing that is being designed in this way, the wider availability of air-conditioned cars and public transport is changing how neighbourhoods and entire cities are being designed. This is because as more people can now travel in the comfort of air-conditioned cars, longer commutes are more viable and common, which impacts on the locations of homes, workplaces, shops and leisure facilities.

This scaling up of air-conditioning is shaping and re-shaping the urban landscape as land is given over to roads and car parks, with social spaces becoming indoor and individualised. Cities are increasingly car, rather than people, orientated. This translates into more roads and fewer paths, pavements and active travel routes. This then creates a self-fulfilling prophecy: the less safe and comfortable it feels to walk and jog, or to travel by bicycle, scooter, or skateboard, the less likely it is that people will choose those options. This apparent absence of demand then makes it easier to justify policies that favour motorists. Likewise, as air-conditioned indoor social and living spaces become more available and more attractive, investment shifts towards them and the supply of – and therefore demand for – outdoor spaces goes into decline. So, as temperatures go up and air-conditioning becomes the norm, people are retreating indoors to the comfort of *coolth* and getting trapped there. Witness the birth of Generation 22°C.

If climate change is forcing us indoors and air-conditioning is keeping us there, Generation 22°C risks becoming consumed by the isolation anxieties described by Sophia al-Maria in Doha. How can we adapt mindfully to climate change and avoid becoming trapped in individualised bubbles of air-conditioning? This isn't just an architectural and engineering challenge; it is a challenge for psychologists, anthropologists and politicians too. Expanded definitions of 'home' might help.

One of the key conclusions the Cooling the Commons researchers have reached is that there might need to be a collective re-think about what 'home' means as climate change makes outdoor living hard to bear. It might be that homes need to be thought of as not just the buildings in which we live, but as the cafés, bars, clubs, streets and parks we frequent, and the neighbourhoods, towns, cities, and countries we identify with.

The 'cool rooms' of Paris are perhaps an early version of an adapted version of 'home' that is fit for extreme heat, but still communal. In the same way that public houses were once frequented not just for the availability of alcohol, but also as places to share the costs of keeping warm, the Paris cool rooms offer an opportunity to socialise around a shared source of *coolth*. Gathering in this way offers economies of scale too, and clear environmental benefits. Ten people sat around one air conditioning unit

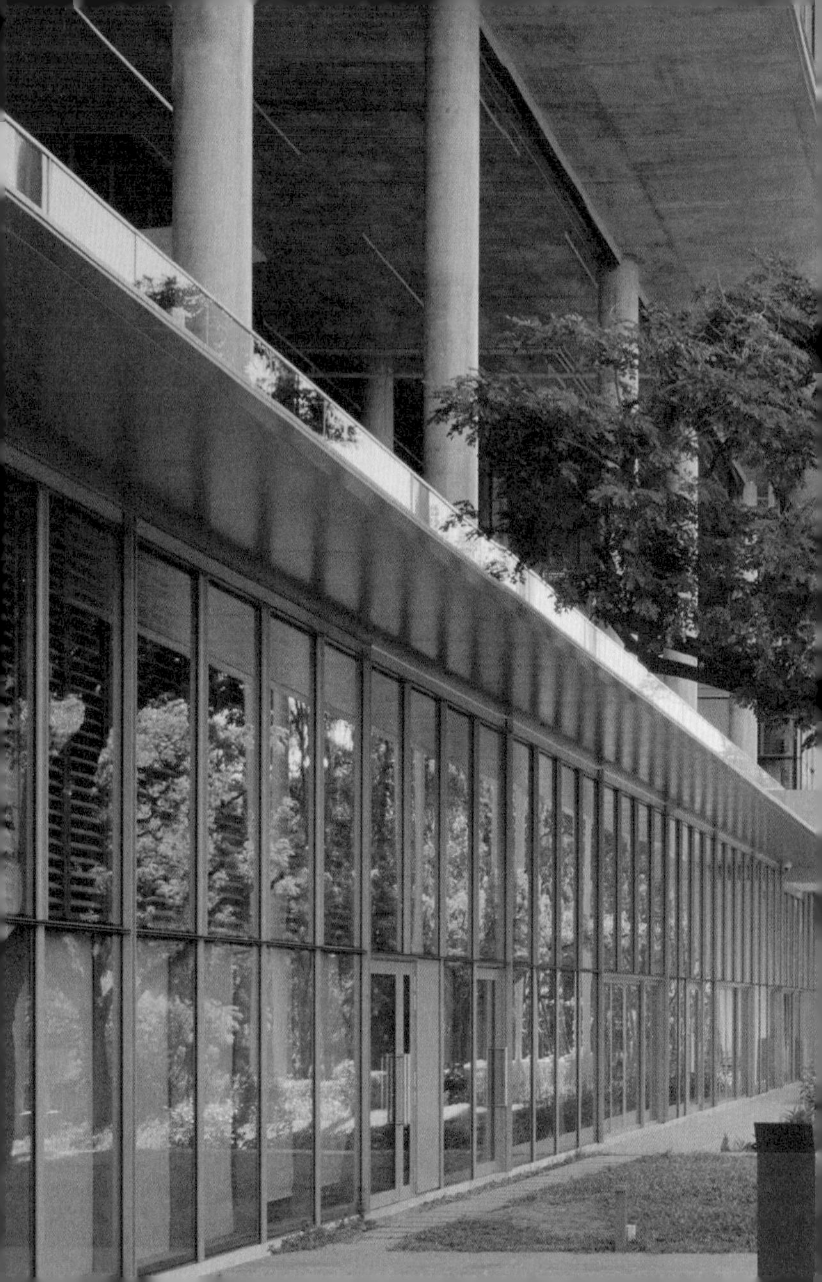

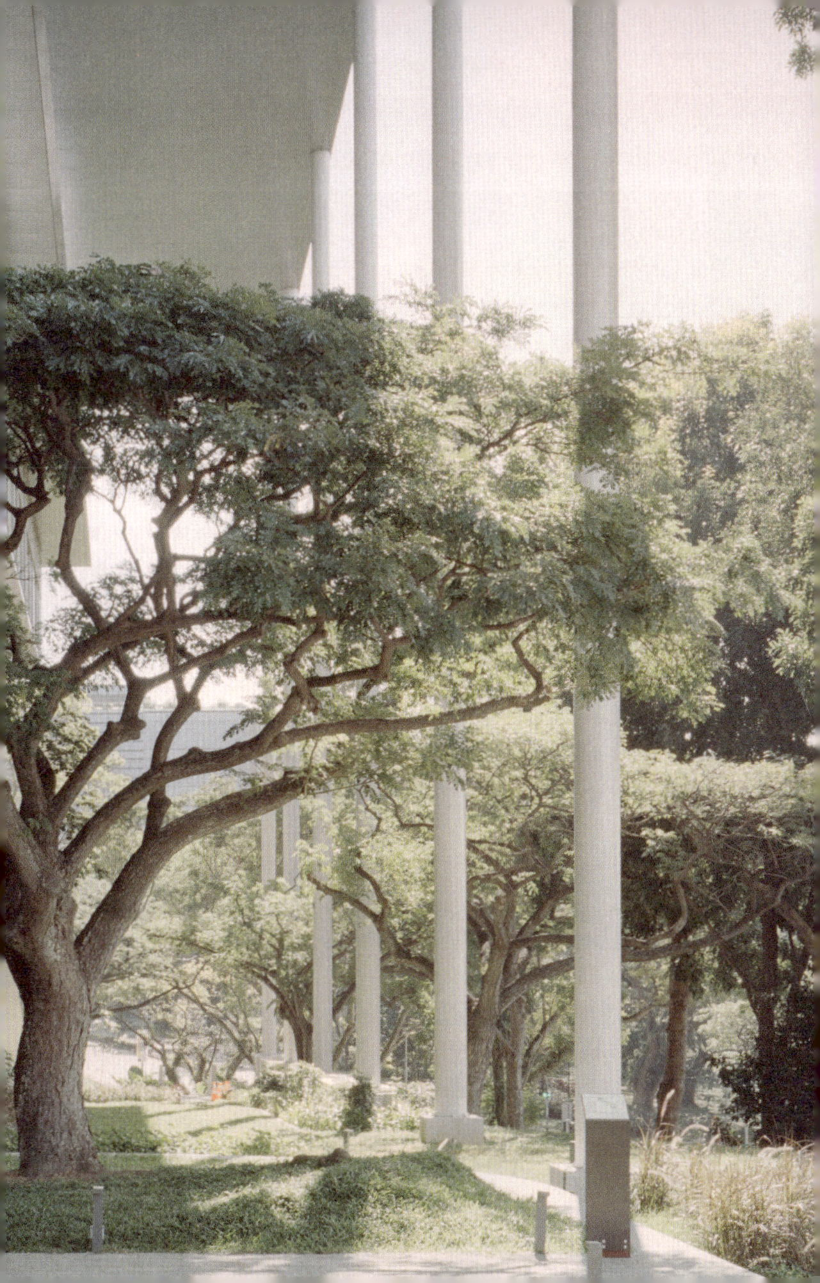

could potentially be using a tenth of the fossil fuel energy that would have been used if they all stayed at home using their own cooling systems.

The inspired School of Design and Environment building at the National University of Singapore shows that it is possible to have a Net Zero, fully air-conditioned and communal indoor spaces in hot countries, but this sort of building is a long way from being the norm around the world. And, even if it were, knocking down millions of existing university (and other) buildings to rebuild them as 'Net Zero' oases of *coolth* isn't really a 'green' option. Retrofitting existing buildings is more practical and more likely in most cases, but this strategy can only get us so far - there are millions (probably billions) of buildings around the world that are performing terribly in sustainability terms.

Generation 22°C need to find ways of keeping cool that are not reliant on air-conditioning. This is important because of the climate impact of fossil fuel powered air-conditioning, but also because *coolth* needs to be affordable. 'Air condition everything' as a policy response has an acute socio-economic dimension. Levels of economic inequality are high in Australia, similar to the UK and USA, which means there are a significant number of people who can't retreat indoors to bathe in fossil fuel powered *coolth*, even if they wanted to. Their options and their experience of life in the city, are nonetheless still being shaped by design principles that favour air-conditioned living. Urban areas that were once pedestrian-friendly are now clogged with polluting vehicles; the liveability of their cities is being incrementally compromised. Cooling the Commons argue that:

> *technical infrastructures of urban cooling that*
> *privilege air-conditioning are threatening the provision of*
> *other infrastructures that afford experiences of coolth, notably*
> *shade, shelter, public water, and places to comfortably*
> *rest and wait while moving about the city.*[45]

These are the issues hot cities all over the world have grappled with for decades – ever since air conditioning was invented – but, as climate change intensifies, so too will the pressure to adapt. Aside from the fact that air conditioning units require enormous amounts of energy and emit HFC refrigerant gases (an extremely potent greenhouse gas), moving towards 100% provision of air conditioning could lead to a variety of negative impacts for people and society. It could exacerbate inequalities that are already reaching excessive levels.

Air-conditioning is a good example of how comparatively wealthy cities like New York, London, Paris, Doha and Sydney can adapt to climate change. They have the financial resources to do so; they can live at a constant 22°C temperature. But, by choosing an *air condition everything* strategy, they risk worsening inequalities in their own cities (as a divide opens up between those who have *coolth* and those who don't) and in the wider world.

A world in which renewable energy is abundant and universally available is still at least a decade away, so if the world's wealthy choose to *air condition everything*, they will be keeping themselves cool, but heating the planet. They can '*air condition everything*', but it is impossible to air condition *everyone*.

And anyway, why would you want to? Who would want to live forever indoors? In a world that is urbanising and increasingly disconnected from nature, reconnection needs to be encouraged, not deterred. We need to go on fulfilling our very human need to spend time outdoors.

# 5. Snow, Grapes, Guns and Dams

These are the early days of climate change adaptation. We are feeling our way; some attempts are better than others. At one end of the spectrum there are brilliant examples of *mindful* and *just* adaptation and, at the other, there are perturbing examples of *mal*-adaptation. Somewhere in between there are multiple brave attempts, well-intentioned efforts, and adaptations that are so instinctual, or so mundane, that those doing them don't even recognise the links to climate change.

People are adapting, businesses are adapting, institutions are adapting, nature is adapting. There are learnings to be taken from all these domains and yet none of what is happening is getting enough attention. This chapter introduces four more examples. The aim is to show how adaptation is being done and how it is being thought about around the world.

It is only by bringing adaptation out into the light that we can critique whether it is being practiced in *just* and *mindful* ways. By doing this, we can hold adapters to account and help them improve the quality of what they are doing. Furthermore, by talking about adaptation, we can highlight the need for it, and therefore increase the supply of resources required to make it happen. This is perhaps the most urgent task; it will help to close the 'Adaptation Gap' (page 11) and enable many people, in many places, to get started on adaptation projects they need. So, just like you might use the stories this book has already covered to spark off a conversation, please use the case studies that follow too - adaptation isn't all about air conditioning everything.

## Don't eat the artificial snow

What do you do if you are a ski resort that has no snow? You can either stay closed and hope the snows come, or you can start making your own. The first time artificial snow was used at a ski resort was in 1952 at New York's Grossinger's Resort[46]. Since then, fake snow has been made and sprayed at ski resorts all over the world. In the European Alps there are resorts that would cease to function without snow cannons.

There are two ways of looking at the *eco* logic of producing fake snow. On the one hand, it is a highly energy intensive process that produces considerable amounts of greenhouse gas emissions. On the other hand, it enables local lovers of skiing and snowboarding to continue their hobby without having to travel, by car or plane, to mountain ranges that still have natural snow.

In some locations, snowfall is now so thin that the volume of artificial snow needed to keep the slopes white is too great to make staying open viable. At the last count, in 2011, researchers identified a total of 186 ski resorts in Italy alone that have been abandoned[47]. That number must have risen by now and hundreds (possibly thousands) of similar resorts in other countries must have gone the same way too. A lack of snow isn't the only cause of this demise: bad management, cheap flights to cheaper ski resorts and changing tastes all contribute. But a higher-than-average temperature rise (a 1°C rise at sea level, equates to a 2°C rise at these higher altitudes) is shortening the season dramatically and these rising temperatures aren't going to reverse any time soon.

Artificial snow isn't the only option for overheated ski resorts. In early 2020, the Prefet du Cantal, a local council in the French

Pyrenees, took the unprecedented step of using helicopters to transport snow from high altitudes down to the slopes at Luchon-Superbagnères[48]. This initiative attracted a lot of criticism, but the Prefet du Cantal's representative, Hervé Pounau, defended the economics of this adaptation strategy. Apparently, the money it cost to hire the helicopter would be recouped in earnings by the ski resort, but he did at least confess that it was *'not very ecological'*. Pounau did also appear to consider his helicopter powered snow shifting as an emergency one-off response: *'It's really exceptional and we won't be doing it again. This time we didn't have a choice.'* If he believes it won't be necessary to do it again, he might be in for a shock.

There is a psychological dimension to this. The helicoptering of snow and the use of snow cannons at ski resorts helps to maintain an impression of normality. It is a form of denial, not denial of climate change as a reality, but a way to make-believe that significant changes to our lives and hobbies aren't yet needed. Whilst there is snow to ski on at Luchon-Superbagnères, everything *must* be alright with the world. And it is not just ski resorts. In Moscow, in December 2019, temperatures were so unseasonably warm that artificial snow was trucked into the city centre to create suitably wintry scenes for the New Year's celebrations. There were even reports of fences being erected to protect the small clumps of snow that had fallen in Red Square[49].

This faking of winter probably did little to assuage concerns of Russians who are experiencing the many impacts of climate breakdown. Traditional family activities like sledging and skating are being lost, cross-country skiing is in decline and, more seriously, across Siberia, prolonged summer heatwaves and

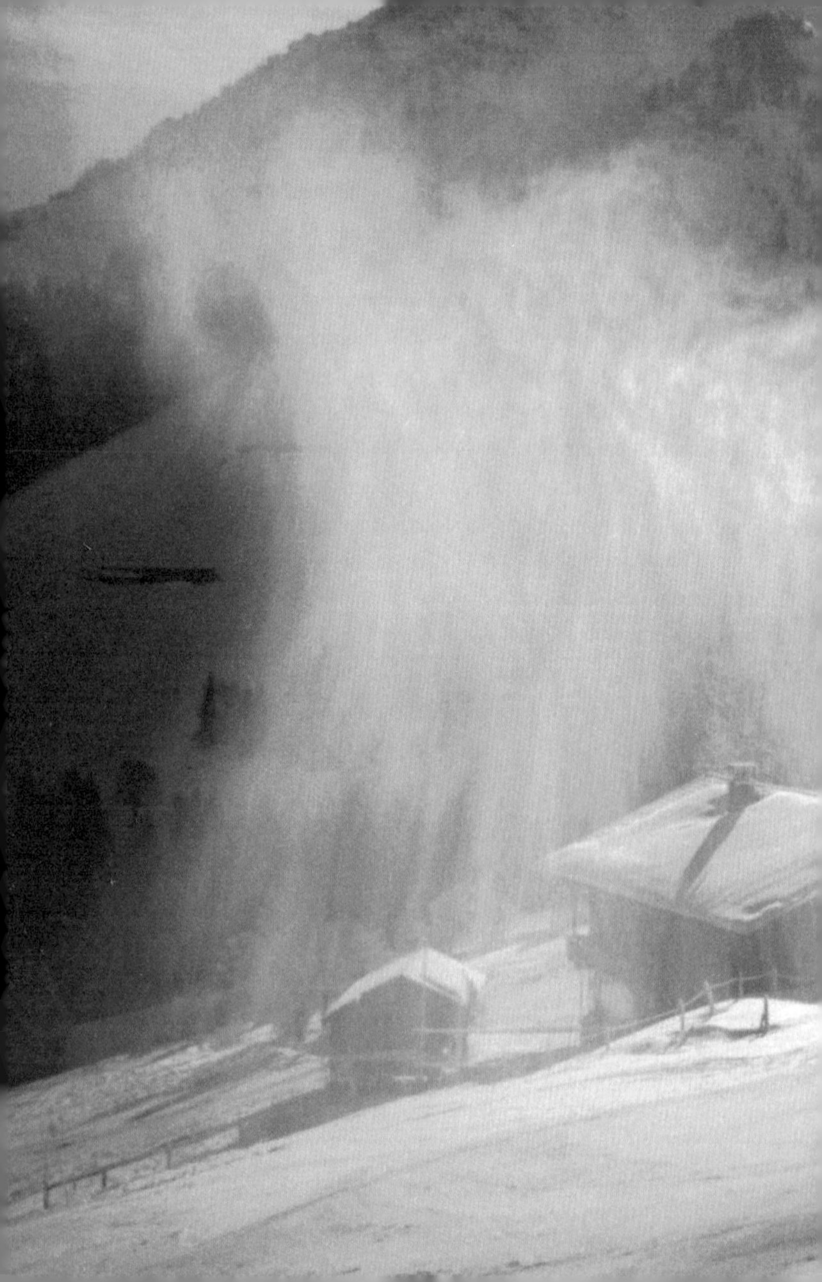

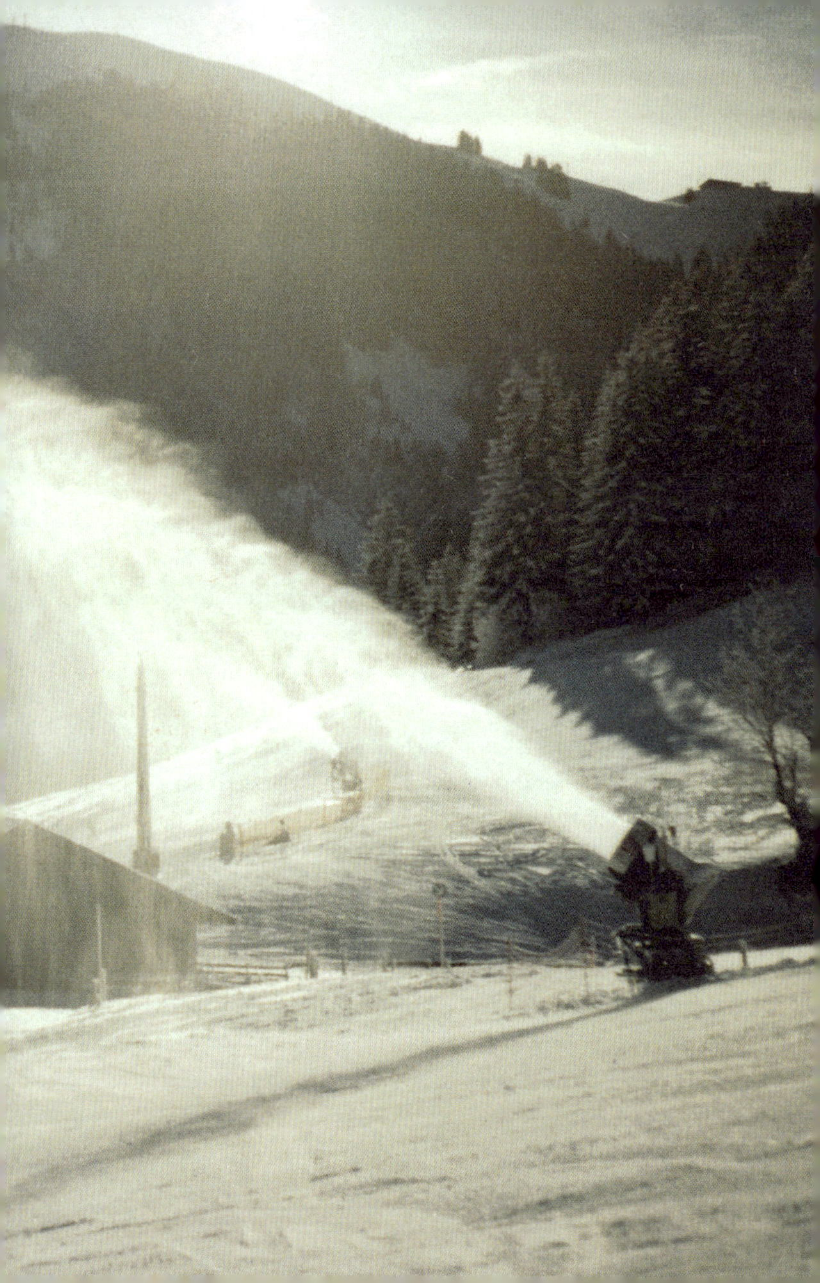

permafrost thaw are becoming impossible to ignore. How long will Moscow's attempts to fake a winter wonderland go on? Will they be enough to distract Muscovites from the climate breakdown they observe all around them? Will it do enough to dampen calls for policies that constrain the use of fossil fuels?

## Grape adaptations

If countries are going to adapt to the impacts of climate breakdown and still thrive, they will have to prioritise their agricultural systems. For many farmers, such as those The Glacier Trust works with in Nepal, a simple adaptation strategy is to switch to producing crops that are better suited to the looming climate conditions. Well, it sounds simple at least. In practice, it is easier said than done. Farmers must learn new skills, acquire new tastes, build new networks, but also let go of old customs and traditions. It requires them to grasp the science of climate change and understand how its impacts might play out.

As well as switching crops, farmers in Nepal also adapt by growing a wider variety of crops. This 'agrobiodiversity' helps their farms and livelihoods to be more resilient. Should an insect pest, attracted to higher altitudes by the warming temperature, destroy one crop, then other crops might survive. This is resilience through agrobiodiversity as an adaptation strategy. These principles are as fundamental to Nepali hill farmers as they are to a vegetable farmer in eastern England, or a cereal grower on the great plains of the USA. But to a wine grower – a viticulturist – a different approach is needed.

Agrobiodiversity is not as attractive to the viticulturist; they are quite happy growing grapes. The world's wine lovers would also rather they didn't switch crop! Sadly though, grapes are very susceptible to increasing temperatures and the freakishness of cold snaps and prolonged droughts or deluges of rain. Research led by Ignacio Morales-Castilla has warned that, with just 2°C of warming, a staggering 56% of current wine growing regions could be lost. At 4°C the loss would be more like an obliteration - 85%[50]. If the wine industry is going to survive, adaptation is an absolute necessity.

Options include relocating to a cooler north facing slope (if you're in the northern hemisphere), to a higher altitude at the same latitude, to a new region, or even to a new country. Champagne companies like Taittinger are already adopting this latter strategy by investing in vineyards that are popping up across the south of England[51]. Parts of England have the chalky soils – and now the temperatures – to grow award winning sparkling wine. It is no doubt more attractive to Taittinger to grab a piece of the faux Champagne market than to be left with a dusty field in the Champagne region. Relocation is an option for the big players in the wine world, but for smaller, family run vineyards, this is almost impossible. They need to look into option two: grow a different variety of grape.

Morales-Castilla has investigated this second strategy and, whilst the findings are promising, it creates the very real prospect that one day soon it will be possible to buy the last ever bottle of cabernet sauvignon grown in Bordeaux. However, as permissions are granted for vineyards to start experimenting with different varieties, it might also soon be possible to buy the first ever

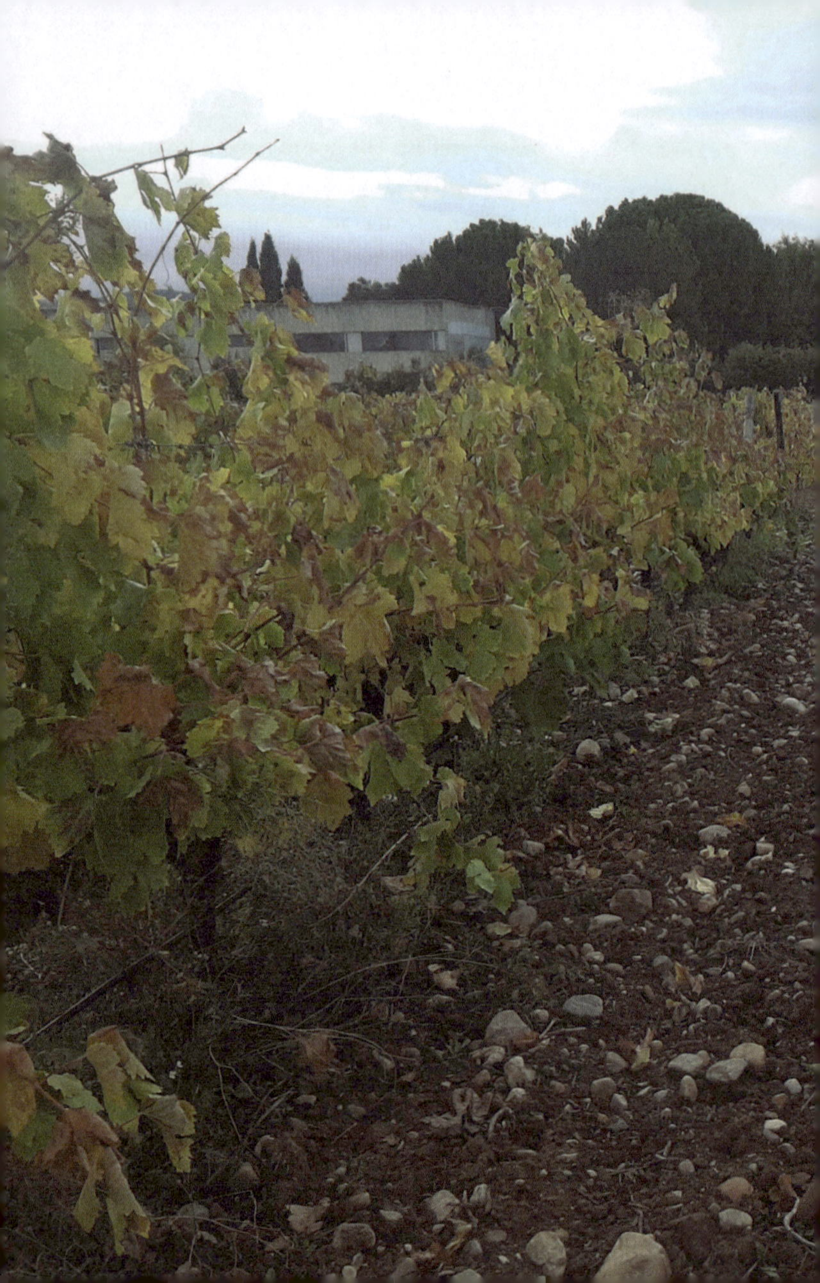

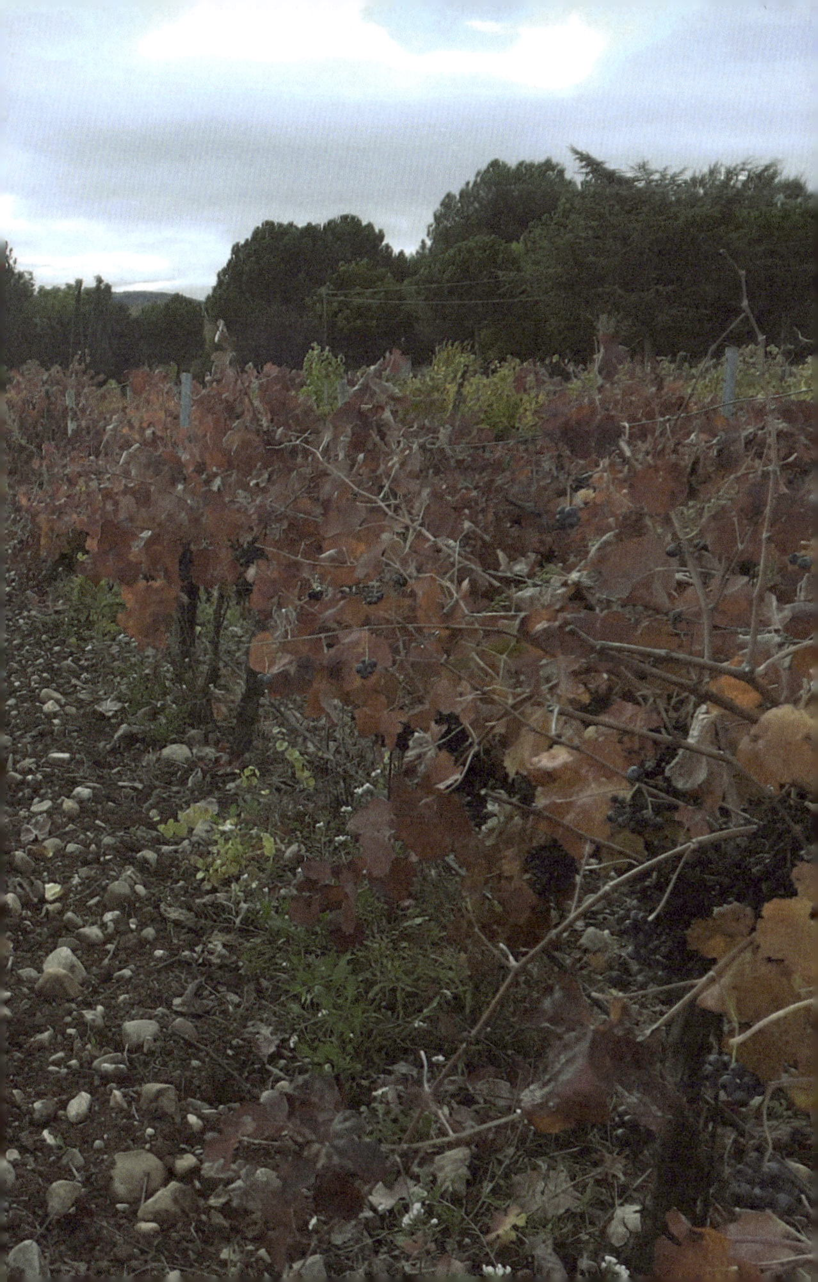

Bordeaux grown glass of Portuguese favourite, Touriga Nacional. However, Morales-Castilla's study shows that even with switches to different varieties, the wine map of the world is still going to change. He predicts that if vineyards do successfully transition to different grapes, 24% of wine growing regions will be lost at 2°C (instead of 56%) and 58% at 4°C (rather than 85%). But this sort of even *relative* success is dependent on agriculturalists receiving accurate projections of temperature increases.

Farmers, from the smallholder to the agricultural industrialist, are used to planning ahead. To plan effectively they need to know what to expect - or at least have a good idea. For viticulturalists to adapt to climate change effectively, they first need to know what they are adapting to. This is where the honesty of leaders and authority figures is crucial. Climate change not only raises average temperatures, it intensifies storms, droughts and floods. In both cases, as a farmer, it is critical to know to what extent. Agriculture is incredibly sensitive to changing weather patterns. Knowing how soon global average temperatures will be, for example, 0.3°C higher than they are today is important. If that rise will take 30 years it might make sense to continue farming the same crop for a while yet, but if it is going to take ten years or less, then it might be wise to change sooner.

Adaptation specialists and campaigners are increasingly concerned that feel-good narratives about climate change are dangerous. They worry that stories about green tech developments and international agreements are overly hyped and therefore paint an overly optimistic picture, which obscures the truth. This problem will be explored in more detail in chapter 11. The main danger lies in the false sense of security *reassuring stories* can generate and the complacency that this breeds. How many agriculturalists are aware of how close the world actually is to going through the 1.5°C, 2°C and 3°C thresholds? Unless they are engaged with specialist climate change literature it is not likely to be many. Those who aren't – the vast majority – are therefore worryingly ignorant about the precariousness of their situation. The more climate sensitive their crops, the more risk they carry.

Farmers also face the unenviable situation of being the targets of climate change protestors, especially livestock farmers. They perceive their livelihoods and cultural heritage to be under threat by those who want to drastically cut the size of the meat, dairy and poultry industries. The wine industry, which is not exactly essential to the survival of the human species, is not immune to these attacks either. Wine production is an intensive process, it uses a lot of water and energy, less than 5% is grown organically, and the final product is shipped in heavy glass bottles all over the world.

Like so many other players across every agricultural sector, the wine industry is under pressure to reduce its wider impact on the climate and ecological crisis. It is trying to do this, while also trying to adapt to the crisis as it unfolds. In the struggle to survive, preserve traditions, and keep our wine glasses topped up, some will embrace the challenge of *just* adaptation, doing everything they can to limit their climate and ecological footprint. Others won't. They will kick and scream their way to a new way of operating, adapting agnostically and with a distinct *'looking out for number one'* attitude.

## Guns don't kill people, climate changes do

When you've got a machine gun around your shoulders and a flock of remote-controlled missile carrying drones flying overhead, it is hard – one would imagine – to remain mindful of your contribution to global climate breakdown. Someone else has hopefully done that thinking for you. However, back in the relative calm of the mess, you might start to reflect on the impacts climate breakdown is having not just on the world, but on you.

A soldier's life is hard enough, global heating is only making it harder. That is why, when a soldier basks in the artificial *coolth* of an air-conditioned mess tent, they might spare a moment to express some gratitude for the adaptations their superiors have made. On the flip side, if their Army's strategists and leaders have been less wary of the changing climate, their soldiers won't thank them - especially when the tent heats up and the water runs out. The military must adapt to climate change too, those who adapt best will steal an advantage.

On a day-to-day basis, in what are often exposed and harsh environments, an army regiment deployed at war needs to be well adapted to its environment. This is why military planners are taking climate change into consideration as they prepare for the coming decades. NATO calls this *'adapting military assets to a hostile physical environment'*[52]. Deployed soldiers need water, food, electricity and, in many locations, air-conditioning. The supply of these essentials is a challenge, especially so in a disaster or war zone. But armies need to go a lot further than this; they are not just responsible for their own. They are most commonly deployed to protect citizens from a threat, or to help them recover from, or survive, a disaster. This means that, in securing the resources that soldiers need in the field, armies have a responsibility to avoid putting further strain on already strained supplies. This responsibility is especially acute in locations where resources are particularly scarce. If they aren't mindful, an army – even a friendly one – can cause immense collateral damage to fragile ecological and social systems, leading to knock-on impacts that harm the very people they are there to serve.

The principle of self-sufficiency helps to mitigate this risk, it always has. However, as physical environments have become more hostile and the energy demands of the modern military rise, the self-sufficiency challenge has grown harder. In recent years, to help improve self-sufficiency, NATO has conducted exercises designed to test technologies that will enable military forces to generate their own electricity and use it more efficiently. The emphasis is on lighter weight and more efficient equipment that can be powered by renewables, rather than diesel. Solar and wind powered generators will, it is hoped, enable militaries to lessen their reliance on fossil fuels. As fossil fuels become more expensive to buy and transport, or more difficult to source in the field, it will become even more advantageous to switch to renewables.

The adaptations the armed forces are making *are* adaptations to climate change, they are quite open about that, this isn't agnostic adaptation. The fact that some of these adaptations happen to involve a transition to renewables is good, but it is important to remember that the goal here is effectiveness, not ecology. The adaptation strategy that is chosen will nearly always be the one judged to be most effective - the army isn't about to switch to greener, but slower, electric tanks. Just like a hospital, police force or fire brigade, the armed forces are not going to compromise their effectiveness on the altar of climate change mitigation. That might sound like a harsh assessment, because many military bodies will – and do – try their best to help, most of them have a recycling policy of some description!

But, when lives are at stake, military effectiveness trumps environmental sustainability every time. NATO is quite blunt about this. As Environment and Smart Energy Officer, Susanne Michaelis, explained to NBC news:

> *NATO's focus is not predominantly on the environmental impact of military activities, but rather on a more efficient use of energy. Put differently, it is about military effectiveness.*[53]

Military forces around the world are clearly thinking about how to adapt to climate and ecological breakdown and making the 'win-win' changes to their operations where they can. However, if we zoom out from the battlefield or the disaster zone, there is a deep-rooted tension that needs to be explored. The military is very firmly knotted up in a truly unpleasant vicious cycle. It is a significant contributor to an increasingly influential root cause of the situations it has to deal with. That root cause? You guessed it: climate and ecological breakdown.

According to Benjamin Neimark and colleagues at Lancaster and Durham universities, the US military alone burns through

270,000 barrels of oil per day and produces 25,000 kilotonnes of carbon dioxide a year. They put those figures in perspective like this: *'If the US military were a country, its fuel usage alone would make it the 47th largest emitter of greenhouse gases in the world, sitting between Peru and Portugal.'*[54] Granted, the USA has by far the biggest military in the world, but add the carbon emissions of all the other army, navy and air forces of the world to the US's and the environmental *'bootprint'* – as Neimark calls it – of the world's military is enormous.

It is not that climate change is disregarded - the Defence departments of most countries, like NATO, list it high among environmental phenomena or issues that pose a security threat. NATO is actively supporting efforts that are *'addressing security challenges emanating from the environment.'* It lists these challenges as including *'extreme weather conditions, depletion of natural resources, pollution and so on - factors that can ultimately lead to disasters, regional tensions and violence.'*[55] Successive Secretary Generals of NATO – Jaap de Hoop Scheffer, Anders Fogh Rasmussen and now Jens Stoltenberg – have all stressed the importance of factoring climate change into medium and long-term defence and military strategy. They understand that it is a deepening crisis.

But to keep things in context, it is important to acknowledge that climate change is not a root cause of *every* military conflict, or *every* natural disaster. It is best understood as an increasingly influential threat multiplier. Climate change events, like droughts that trigger crop failures, can be enough to tip a precarious but peaceful situation over into a volatile and hostile crisis. Recent civil wars in central Africa, for example, are strongly linked to the drying up of Lake Chad and widespread loss of fertile lands, and many have argued that climate change plays a role in creating unrest and conflict in the Middle East[56]. Climate change is also increasing the frequency and intensity of natural disasters that military forces are called to respond to. They are regularly present to deal with severe flooding, landslides, storms and forest fires, and they don't turn up on their bicycles. Imagine having to deal with the fallout of a problem using equipment that exacerbates that very same problem? That is what is happening, it is a very tangled mess.

The NATO Secretary-General understands all this. He said in a speech in 2019: *'Climate change has security implications. It can force people to move, change the way we live, where we live, and so on, and of course that can fuel conflicts.'*[57] But he doesn't view NATO as the body to solve climate change, he charges the UN with that responsibility. So, although he is clearly concerned about climate change and supports efforts to mitigate it – including those being made by military bodies – the prevailing attitude seems to be that if, in dealing with the security implications of climate change, NATO and its members must cause more climate change, then so be it. One thing therefore seems certain: the military will keep on contributing, and responding, to climate change for a good few decades yet. It will keep itself busy.

## Dam good idea?

Climate change has geopolitical implications too. Previously impassable shipping routes through the polar regions are opening up, it is becoming possible to grow crops at higher altitudes and latitudes, and fishing grounds are shifting gradually across the oceans. As lands and seas become more valuable, the scramble for control of them heats up. Businesses and nation states want access and control. Similarly, as the energy transition speeds up, the demand for locations to site green energy tech is also on the increase.

Political leaders at all levels of Government are therefore adapting their geopolitical ambitions to the emerging climate reality. Some are spying new opportunities, others are yearning to re-locate as they suffer losses and damage to homes and infrastructure. Various degrees of soft and hard power are being used to achieve these goals. There will be disputes over rights, access, territory, and borders; it won't always be possible to avoid acts of aggression. The potential impacts of climate change are so great that it looks set to be something that shapes the geopolitics of the future, rather than something that countries must adapt their usual geopolitical ambitions to.

---

Here is a thought experiment: what would be the geopolitical implications of enclosing the English Channel and North Sea so that it becomes a giant sea water lake with a sea level that is artificially lower than the surrounding Atlantic Ocean? That is not a question anyone would ever have expected to write, but it is a question that needs asking if the proposals of two Dutch oceanographers are to be taken seriously.

In 2020, Sjoerd Groeskamp and Joakim Kjellsson scoped out a (very) big idea[58]. They tried to calculate the cost and feasibility of building two dykes or dams, one stretching from the tip of north-east Scotland to the west coast of Norway, the other from Cornwall to Brittany. These dams would protect the east coast of England and Scotland, along with the North Sea coastlines of France, Belgium, Holland, Germany, Denmark and Norway from sea level rise. It would presumably spare the Swedish, Finnish and Baltic coastlines too. But not Ireland, nor the western or northern coasts of Great Britain... nor western France, nor Spain, nor Portugal.

Groeskamp and Kjellsson's proposal for a 'NEED' (Northern European Enclosure Dam) was, they argued, technically possible and affordable (i.e. it would be cheaper to build a NEED than to retreat from the shorelines). They were not, however, being serious. The paper was conceived and intended as *'a warning of the immensity of the problem hanging over our heads'*[59]. In that respect it works. The danger, of course, is that one day someone, somewhere, *might* take NEED seriously. Milton Friedman is famous for saying: *'Only a crisis – actual or perceived – produces real change. When that crisis occurs, the actions that are taken depend on the ideas that are lying around'*[60] We had better hope a better idea comes along soon because populist politicians, of the sort Europe now seems to be electing, are fond of €500bn mega-projects they can plaster their name all over.

The problem with 'kite flying' an idea like NEED is that it is so outlandish that more serious big ideas seem less mega in comparison. Some of the real projects under consideration or construction around the world would – by any other comparison – look absolutely epic. Next to NEED, however, they look quite restrained. In an attempt to hold back the rising seas and superstorms, humanity looks set to unload enormous volumes of aggregate and concrete onto its shorelines over the coming decades. There are serious proposals for a sea wall that stretches from northern France to Denmark, but the most high-profile and advanced plan to adapt to rising sea levels can be found in Miami, Florida. The Miami plan, not NEED, is the sort of mega adaptation project that others should be judged against.

Miami Dade County is low-lying, densely populated, and subject to tropical storms. It is home to 2.8 million people and is one of the USA's most prosperous tourist and trade centres. It is vitally important to the nation economically and culturally, but it is vulnerable. Furthermore, continued sea level rise, combined with the predicted increase in frequency and intensity of hurricanes, makes it more vulnerable to climate change each year. There is much debate over what is needed, where it should go, and who should be compensated.

Residents, engineers, journalists, and policymakers are debating plans that the US Army Engineer Corps have been asked to propose. The aim is to reduce both the occurrence of flooding and the damage done by flooding that cannot be avoided. Proposals include a series of walls, totalling six miles, that are interspersed with temporary storm barriers that can be moved into place before a storm hits. Hundreds of houses and buildings will need to be demolished to build the walls - which will, in places, be so tall (up to four metres high) that they will block the treasured ocean views of some of Florida's wealthiest homeowners. Presumably, however, losing a view is still preferable to being on the ocean side (i.e., the wrong side) of the walls, which some houses and businesses will be.

Also proposed are plans to fully floodproof critical structures like hospitals, fire and police stations and water treatment plants. Remarkably, planners are also proposing to elevate over 2,000 residential homes, raising the ground floor by three metres or more. A further 4,000 buildings that are too big to elevate will be flood-proofed by fitting watertight doors and flood barriers. These proposals are estimated to cost $5bn, and the local economic, social and environmental impacts are extremely complex - the May 2020 feasibility study ran to 443 pages[61]. Adaptation, at this scale, is an extraordinarily convoluted undertaking; work is due to start in 2023.

Campaigners from across the political spectrum are resisting or promoting various parts of the plan in a predictably diverse way. Some argue that the plans are too costly (an earlier $8bn version has already been abandoned), others are saying the plans aren't sufficient and worry that a lot of money will be spent on megastructures that turn out not to be fit for purpose[62]. But it is the socio-economic implications of the plans that require the most scrutiny. We often think of geopolitics as global phenomenon, but local geopolitics are equally complicated and fraught. There is a lot up for grabs: the whole of Florida will be adapting in the years ahead, lives and landscapes look set to change beyond recognition; similarly vulnerable places will observe with great interest.

Miami is the USA's ground zero for a new phenomenon dubbed *'climate gentrification'*, a process that is starting to spread across the US. In Miami, traditionally wealthy neighbourhoods like South Beach and Brickell are low-lying and either soon to be, or already, suffering from flooding. Residents are keen to move to higher land. Historically many of Miami's poorer neighbourhoods like Little Haiti, Liberty City and Allapattah are inland and at slightly higher altitudes. Their height and distance from the coast used to make them *less* desirable, but it is now what makes them *more* desirable. Property prices are therefore on the up, landlords are pushing rental prices in the same direction and the wealthy are elbowing their way in. Those on the wrong end of this gentrification process – who called Little Haiti, Liberty City and Allapattah home for several generations – are being gradually squeezed out.

It is worth remembering that hurricanes and rising sea levels are only two of the climate change impacts that Miami Dade needs to adapt to. It can be a very hot place, in the years to come it will suffer prolonged droughts and blistering heatwaves, billions will be spent to secure water supplies, and billions more on air conditioning. And so, as it adapts to climate change, will Florida be mindful of the climate and ecological impacts of the strategies it adopts? Current signs are not good. Nature-based adaptations were looked at as part of the feasibility study, but largely deemed to be unworkable. There will be no restoration of mangroves or submerged aquatic vegetation, nor any development of 'living shorelines' or coral reefs. Instead, the plans for Miami involve a lot of construction and a lot of concrete. In short, by adapting to climate change, Miami will contribute to climate change. It won't be the only city.

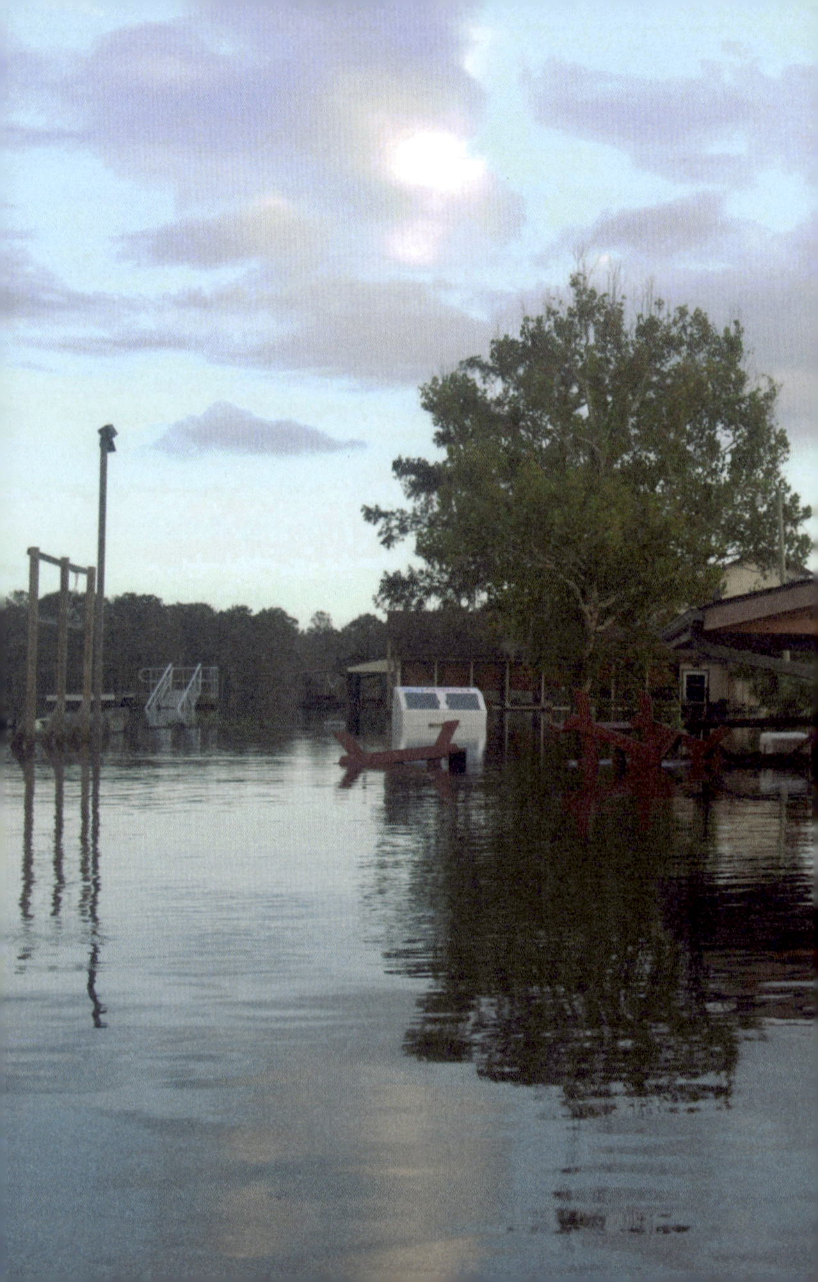

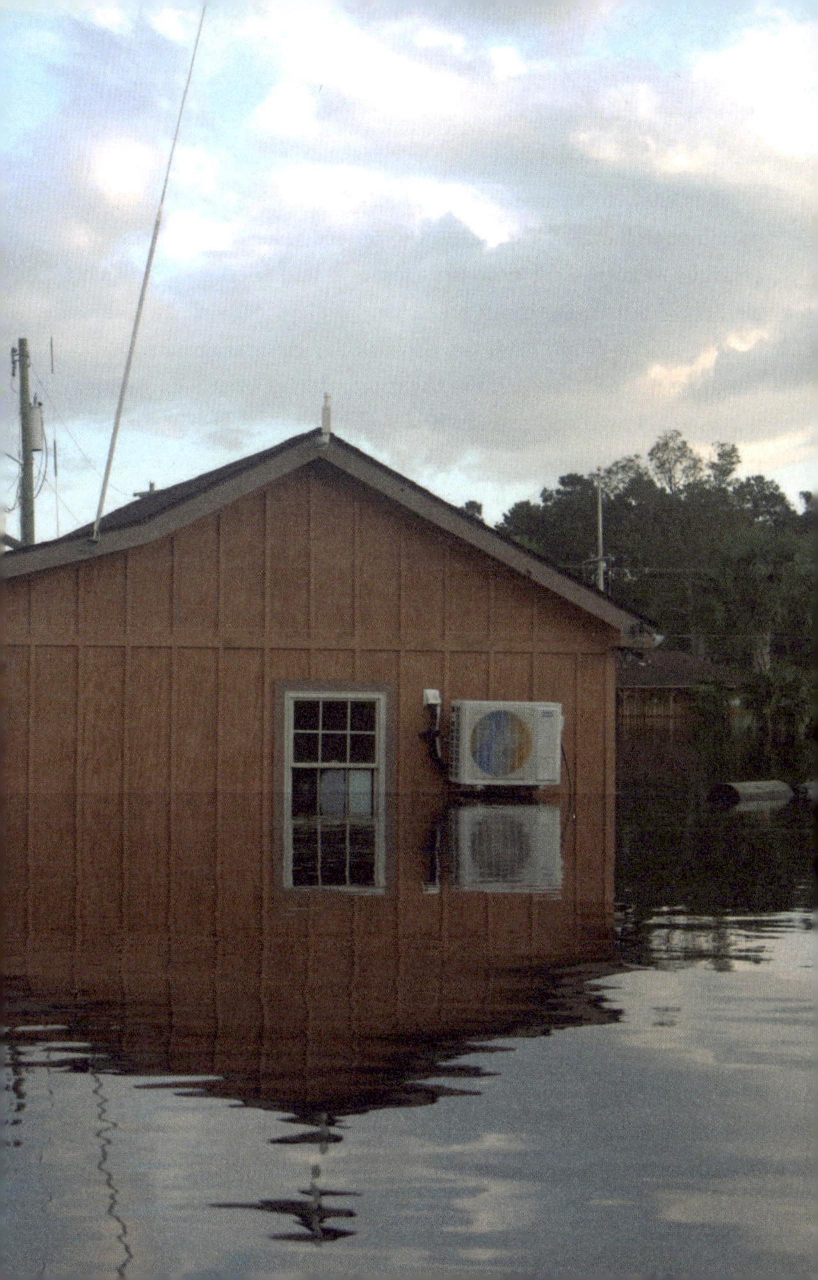

## Adaptation First

It is easy to become frustrated at the apparent lack of action on climate change. Despite the widespread levels of concern and growing awareness of how little time is left, inertia reigns. Or at least, that is how things seem. On an airplane, shortly before take-off, the in-flight team run through the safety announcements. Towards the end, they point to the compartment where the oxygen masks are stowed and explain how and when to use them. At this point there is always a specific piece of advice for those flying with family or friends: *'be sure to put your own mask on before helping others'*. This is what is happening in the business world. There is far more action on climate change than businesses usually get credit for, but it is mostly at the *'put your own mask on'* stage. Very few have got onto *'helping others.'* Even fewer are *thinking of others*.

Shareholders, CEOs and Directors are in the midst of a gradual collective awakening. The proximity and scale of the climate emergency is slowly (it is taking decades, not years) coming into view. As they wake up and blearily start to focus, they take the most natural and sensible first course of action: they prepare *themselves* for what lies ahead. They are taking their sleeping masks off and putting their oxygen masks on.

There are so many examples of businesses who are taking action on climate change in this way, enough to fill an entire book, certainly enough to fill a PDF download. There is one called 'The Business Case for Responsible Corporate Adaptation: Strengthening Private Sector and Community Resilience' by the UN Global Compact[63]. It weighs in at 96 pages and is the sort of document that goes almost instantaneously out of date. What it illustrates, however, is that the adaptations businesses are making are not being deliberately hidden. They can be found, they are just not being talked about - they are in the shadows.

One of the best climate change writers currently working, Eric Holthaus - author of the excellent 'The Phoenix' newsletter[64] - wrote an article in 2020 that neatly summarises three climate change eras[65]. There's the *'climate stability'* era, which [spoiler alert] has passed; the era of *'irreversible weather disasters'* (which we seem to be very much in); and the *'post-tipping points'* era, that desperately needs to be avoided. As the *'irreversible weather disasters'* era gets worse, it is going to trigger an explosion of activity in the business world.

Pretty soon, businesses will realise that their supply chains, customer base, factories, shops, and offices are at risk of collapse because of climate change. These laggards, regretting that they put off attending all those dry sounding adaptation webinars they could be tuning into now, will be madly scrambling to adapt to the *irreversible weather disasters* era and the chaos it will cause. This stampede will make a handful of adaptation consultants very rich, and the scale of changes they recommend will make today's adaptations to COVID-19 look quaint. As ever, the concern is that in the rush to adapt, the strategies adopted will be ad hoc at best, and dangerously *mal*adaptative at worst. Instead of calmly putting their own masks on before helping others, businesses will be lunging desperately for the oxygen and trampling all over workers' rights and environmental protections, causing all manner of collateral damage. They will be more of a hindrance than a help to everyone else - and themselves. It is why framings like

#RaceToZero and the newly launched sister #RaceToResilience are a little problematic.

To conclude this chapter, it is worth turning to the advice that exists for businesses looking to avoid *mal*adaptation. A 2014 paper by Alexandre Magnan lists eleven guidelines for avoiding *mal*adaptation[66]. He divides them into three categories - environmental, sociocultural, and economic. They are specific to coastal areas, but can be adapted to apply in all contexts:-

### Avoiding environmental maladaptation

1. Avoid degradation that causes negative effects *in situ*. (e.g. don't destroy sand dunes that would ultimately be a more effective protection against sea level rise than a concrete wall).
2. Avoid displacing pressures onto other environments (neighbouring areas or areas that are connected ecologically or socio-economically).
3. Support the protective role of ecosystems against current and future climate-related hazards. (e.g. conserve tropical rainforests and mangroves).
4. Integrate uncertainties concerning climate change impacts and the reaction of ecosystems (ensure your strategy is flexible enough to cope with worst-case scenarios).
5. Set the primary purpose as being to promote adaptation to climate-related changes rather than to reduce greenhouse gas emissions (ideally adaptation measures will be carbon neutral, but they do still need to work; so a balance needs to be struck).

## Avoiding sociocultural maladaptation

6. Start from local social characteristics and cultural values that could have an influence on risks and environmental dynamics (some adaptation strategies will impact more than others; make sure they don't clash with the expectations the community has about its own adaptation plans).
7. Consider and develop local skills and knowledge related to climate-related hazards and the environment (the local community will have varying degrees of awareness and knowledge on the climate risks faced, engage them in the process).
8. Call on new skills that the community is capable of acquiring (and help them acquire them, support efforts at environmental education).

## Avoiding economic maladaptation

9. Promote the reduction of socio-economic inequalities (and make sure your adaptation strategy doesn't exacerbate existing inequalities).
10. Support the relative diversification of economic and/or subsistence activities (e.g. farm a more diverse range of crops).
11. Integrate any potential changes in economic and subsistence activities resulting from climate change (i.e. don't rely on suppliers who are highly vulnerable to climate change).

Businesses will adapt or die. For many, that is the stark reality of what lies ahead. So when assessing how much action a business is taking on climate change, it pays to examine more than just their emissions reductions efforts. In the shadows of the slick presentations, the greenwashing and the sloganeering are their adaptation strategies. Some are adapting responsibly and others aren't, but - for better or worse - they don't talk about it much. And why would they? Who's asking?

# 6. Adaptation in the wild

The purpose of this book is not to convince anyone that climate change is real, or serious, or an existential threat. If you have read this far, you probably don't need to be persuaded. But if you ever find yourself needing to convince someone else that climate change is real (deniers do still exist), what follows are a few examples of how the natural world is adapting. What better proof is there? The plants and animals aren't in on the hoax are they?

It is always a little bit tempting to have a go at the deniers, but that is not why this chapter has been included. What follows are some of the examples of the adaptations being made by plants and animals around the world. Biologists have identified three main trends: (i) species migration, (ii) species downsizing and (iii) changing phenology (a change in the timing of annually recurrent biological events such as flowering and nest building). These trends are being monitored closely by conservationists as they seek to protect species and habitats where climate change is compounding multiple other challenges. There may also be lessons for humans, things to copy and mimic:-

> *Biomimicry is a practice that learns from and mimics the strategies found in nature to solve human design challenges — and find hope along the way.*[67]

What can the natural world teach us about adaptation?

## This ice sheet ain't big enough for the both of us

In 2015, a team of scientists for the US Geological Survey confirmed something that had long been suspected. Polar bears are adapting to warming in the Arctic Circle by migrating steadily north[68]. The movement northwards is a creep and detected by analysing their directional gene flow over time. It is the polar bear version of an agnostic managed retreat from the front line of climate breakdown. But they might not be able to out-run their fate, because the Arctic might experience summers that are 100% ice free as soon as 2034[69]. The polar bears' northern creep might soon turn into a sprint to the worst kind of finish line.

Interestingly, even though the overall trend is a northwards migration, polar bears are adapting in far more headline grabbing ways too - by heading south. In 2019, on the Novaya Zemlya archipelago in northern Russia, photos of a sleuth of polar bears started to emerge. The bears were in the small town of Belushya Guba. They were scavenging for food in bins, apartment blocks and school playgrounds. The local authorities had to fire warning shots, erect fences and warn residents that over 50 hungry wild bears were in their town.

Top of the food chain predators like polar bears need a sizeable territory to call home. If their natural habitat starts to shrink a *'this town ain't big enough for the both of us'* situation arises. A habitat that was once large enough to support two sleuths of polar bears shrinks to the point where it is only big enough for one. At that point, the weakest sleuth is forced out.

It was climate disruption that forced polar bears south to

Belushya Guba, they were seeking refuge there. And like so many other climate refugees, they weren't given a warm welcome. As the Guardian's environment editor, Jonathan Watts pointed out they were talked about as *'invaders'* that needed to be *'deported'*[70]. In reality, they are victims of climate and ecological breakdown and desperately trying to adapt.

## Camels in the crosshairs

In southern Australia, another wild animal has been forced into human settlements to scavenge for survival. The impacts of climate change have become so acute in parts of Australia that water supplies are running very low, sometimes dry, in the summer months. The scarcity is felt by humans, but also by animals like camels that live wild in the bush. The camels are adapting by entering onto farms, into villages and small towns in search of water. They gnaw at water pipes, tanks and air-conditioning units. The reception they receive is hostile to say the least; water is in short supply and humans are not in the mood to share it.

In January 2020, in the midst of the forest fires that engulfed huge parts of Australia, Aboriginal elders gave permission for a cull. Professional snipers flew over the desert, shooting camels from the sky. Echoing the language used against the polar bears in Belushya Guba, several media reports described the camels as *'dangerous competitors'*[71] and much was made of the fact that camels are non-native *'introduced pests'*[72].

Curiously, one of the justifications given for the cull was climate change related. The 10,000 camels in the crosshairs were framed as a source of greenhouse gas emissions. It is true, they are; and there are an estimated one million camels in Australia. They release methane into the atmosphere just like Australia's 25 million head of cattle do; but shooting camels to mitigate climate change seems to be a rather unfair piece of victim blaming. There are perhaps other strategies Australia could prioritise if they truly want to lower their contribution to climate breakdown.

## Icequake!

When you pour a warm drink into a glass filled with ice it makes a quite dramatic crackling sound. The same thing happens at sea when icebergs drift into warmer water and break up. A team of oceanographers at Oregon State University[73] set out to measure the noise caused by these *'icequakes'*.

Their findings surprised them. The icebergs generate a lot of noise - far more than they expected. The noise doesn't come from the collisions with rocks or the seabed, it is purely the noise an iceberg makes as it disintegrates.

As temperatures rise and ice caps melt, icequakes will become more common and the noise they generate is not just heard locally. Hydrophones as far away as the equator were detecting noise made by the icequakes studied by the Oregon team. Ocean animals, who rely on sounds to navigate, hunt and breed, will need to adapt.

More straightforward adaptations have been observed in the oceans too. As temperatures warm, fish populations are migrating in search of colder waters. David Wallace-Wells reports how Mackerel have headed ever further north in the eastern half of the Atlantic, and Flounder in the western half, are now at least

250 miles north of their previous habitat[74].

## Small is beautiful... and adaptable

Before getting too carried away with melting ice caps and the adaptation strategies of the world's charismatic megafauna, it is worth highlighting the second trend that climate change is helping to drive in the animal world: species downsizing.

Researchers from the University of Southampton are predicting an overall downsizing of birds and mammals over the next century[75]. Small animals, such as birds and rodents, are fast lived, highly fertile and have a more generalised diet, which makes them more adaptable and resilient to environmental change. They therefore argue that larger animals, like camels and polar bears, but also rhinoceros, elephants and eagles, will be slower in adapting to climate and ecological breakdown.

The rate at which this downsizing is predicted to occur – and it is already underway – is astounding. Over the last 130,000 years the median average body size of mammals has decreased by just 14%. The team at Southampton predict that over the next 100 years, the average body size of mammals will decrease by a further 25%.

## Adder-daptation

The third trend – phenological change – is where things get even more interesting. In March 2019, the Amphibian and Reptile Groups of the UK reported that Adders were, for the first time, active all year round[76]. An unseasonably warm February had brought some out of hibernation early and others have previously been recorded as being active in November, December, and January. This is not so much an adaptation, but a consequence of the milder winters climate change is causing.

Waking up early is not a smart adaptation for the adder. Years of habitat loss, disturbance and inbreeding already means they are in decline and vulnerable. Coming out of hibernation and the security of their winter nest or burrow exposes the adder to predators and any cold snaps that might follow a spell of warm weather.

Climate change is leading to shorter hibernation periods for many animals. Some wake earlier like the adder, but others settle down later because the warmer temperatures mean that food is available for much longer into the winter months.

Articles by Renne Cho[77] from Columbia University highlight the impact climate change is having on animals and insects. She notes how several are apparently adapting. Migration is the most common strategy (one that humans are certainly mimicking), but Cho points out how birds are also laying eggs earlier to align with when the insects are available to feed their chicks.

## Plastic coral

There is a fourth related trend, phenotypic plasticity (the ability of an organism to change its behaviour, development and physical features during its own lifetime - rather than through evolution). In the oceans, some species of coral with phenotypic plasticity have been found to adapt effectively to warming waters. Corals contain a special photosynthetic algae, zooxanthellae, that supplies them with sugars for growth. When corals come under heat stress, they expel this life sustaining algae from their bodies, but the more they expel the less they have available to them as food. A laboratory study of corals from the coast off American Samoa[78] observed how some coral species have used their phenotypic plasticity to expel less algae than they normally would in warm water. The result being that they maintain their healthy, colourful state, rather than bleaching (and dying).

This ability to adapt is down to a species' potential for epigenetics - the process of a gene turning on or off depending on external environmental factors. Some species of coral can do it as an adaptation to warning ocean temperatures and researchers are discovering an increasingly diverse range of other animals can do it too. Renne Cho describes how animals as diverse as Guinea Pigs, Winter Skate fish, Voles and Turtles all display signs of epigenetics[79].

## 7. Survival of the friendliest

With a few honourable exceptions, this book has mainly covered the improvised and shadowy sides of adaptation. These stories of *mal*adaptation highlight the pitfalls of the strategies being pursued. They also show us why adaptation needs to be done in smart ways, and why holding powerful actors to account as they adapt is such a priority. There are, however, brighter stories and signs of hope. As adaptation needs spring up around the world, examples of mindful, *just*, effective, and scalable strategies are starting to emerge. Shining a light on these, bringing them out of the shadows, is just as important as exposing the multiple *mal*adaptions that give adaptation a bad name.

The good news is there are plenty of excellent examples. Buried in the reports, the academic papers, the blogs, and the books are the case studies that show the way to a better future. These adaptations are the ones that can inspire and educate us. They speak of humans acting in *bigger-than-self* rather than *selfish* ways - which, on the whole, is what most people are predisposed to do (see page 103).

---

In the shadow of the seemingly incessant flow of bad news stories that we doom-scroll our way through every day are a multitude of understated stories of human goodness. They are the less newsworthy stories that we only hear occasionally, most never get an airing. We are deaf to them *because* they are so common; human beings are forever doing things for each other and the planet. There are so many stories of goodness that only the most

In his 2020 book, Humankind[80], Dutch historian, Rutger Bregman, reminds us of a foundational truth about our fellow humans. He does this by directly challenging a powerful myth about human nature that has been allowed to grow. This myth is that below a thin veneer of apparent order, is a (non) society of selfish, corrupted, vicious individuals who are desperate to break loose from their supposed shackles. Bregman argues that while this myth is powerful and has a strong grip on our imaginations, ultimately it is false. The majority of us aren't self-obsessed, callous profit maximisers. We are kind, we are friendly, we respond to crises with compassion not competitiveness. We cooperate, we give, we seek justice, we protect nature, we stand-up for what is right and *together* we thrive. Human history is not so much a story of survival of the fittest, it is survival of the friendliest[81]. Sure, there are narcissistic, image obsessed and genuinely evil people, and they regularly make the headlines, but they are a tiny minority. In reality, as Bregman puts it, *'most people, deep down, are decent'*.

So, to be clear, it isn't my contention that the *mal*adaptations covered in this book are *'mal'* because the people responsible for them are fundamentally selfish, or uncaring about the knock-on effects of what they are doing. While a few people or businesses are looking after number one in the ways they are adapting, most are *mal*adaptating for other reasons – they are lacking resources, they are aiming at short term survival, they have been given perverse incentives, they are forced into compromises, they are following advice from people they trust – they aren't bad people.

extreme ones cut through. For every Captain Tom Moore, who raised £33 million in 2020 for the NHS, there are 33 million fundraisers who, between them, are raising ten times that total in almost complete anonymity. We don't hear about them, it would be impossible for the news organisations to cover them all, so they don't cover any, which is precisely why it is so easy to underappreciate how much good is actually happening. Good news doesn't cause the same stir as bad news.

We could all, of course, do more good, and most of us would like to, but sometimes we need to be shown the way. We need to feel that it is normal to do acts of goodness and, for the more complicated things that need doing, we need to learn how to do them. This is as true for *mindful* adaptation as it is for anything else. Presented here are four stories that will give you hope. They tell of a commitment to *mindful* and *just* approaches to adaptation. If these stories are told and told well, they will provide more than just hope. These stories have the power to guide and instruct the adaptation journeys of people, communities, institutions, cities, businesses, and governments. These are the early adapters we can be proud of. Breathe them in, they are the antidote and the cure.

## Drinking Fog

The world is littered with the broken and discarded remains of well-meaning technological solutions to social and environmental problems. On far too many occasions, NGOs have imposed their ideas onto unsuspecting people in clumsy and rushed ways. Quite often it isn't the technology itself that is at fault, but rather the failure to establish in advance – and with the community – if the technology is the best fit for the unique circumstances of the society and environment it is landing in. It sounds a little silly to say it, but if a community doesn't see the value of the adaptation strategy it is being told to adopt, it is probably not the right strategy for that community. This can be frustrating because climate change adaptation is nothing if not urgent and people want to help. But, it doesn't matter how innovative or proven a strategy or piece of technology is, if it doesn't make sense it won't take off, it'll go to rust.

In southwestern Morocco, in Aït Baâmrane, on the edge of the rapidly expanding Sahara Desert, a local NGO called Dar Si Hmad has developed an innovative climate change adaptation project[82]. It is now one of the most celebrated in the world. It involves 'Fog Harvesting', a simple technology that, crucially, creates more than just one positive outcome for the villages it serves. 'Fog harvesting', or 'catching', isn't a new invention, archaeologists in Israel, Egypt and South America have found the remains of stone structures built to capture evaporated water that date back to ancient civilisations. But the water stress brought on by climate change has led to a revival, fog harvesting is back.

The technology is relatively simple. A net, or mesh, is hung vertically from a frame that is anchored in a fog prone location. Fog catchers are no bigger than $3m^2$ and are most often erected on hillsides, usually with several others. The net catches the fog as it rolls uphill, water droplets then form and trickle down into a gutter that feeds a pipe to take the water to nearby homes and fields. Well maintained and advanced systems can turn fog straight into drinking water; drinking fog.

Climate change is hitting southern Morocco hard, droughts are intensifying, rainfall is sporadic, and groundwater is receding. The lack of water, combined with bouts of intense heat and the encroaching desert, is making agriculture very difficult. As a result, men and women (but mostly men) are emigrating from rural areas to seek work and money to send home to the families they leave behind. Those who remain are heavily reliant on the women of the household. These mothers and daughters have many responsibilities, but number one is to collect water. This story is a common one across Morocco and around the world, the villages of the Aït Baâmrane region are no exception. Even before being hit by climate disruption, women were spending a lot of time and energy on the never-ending task of supplying their families with the ten litres per day of water they need. As climate change intensified, rainfall declined, and local wells started to dry up. Women were forced to walk further and further in search of a water source. Some were spending over three hours a day collecting water, it was getting intolerable.

The fog harvesting system built by Dar Si Hmad is seriously impressive. On hillsides above the villages, they installed 600m$^2$ of nets to catch the morning fog. The harvested water then flows down 10,000 metres of piping to seven reservoirs (that between them have a storage capacity of 539m$^3$) and on down to 400 people, most of them women, in 52 homes across five villages. This system supplies homes with more than enough water, around 12 litres per day, and is paid for via pre-paid water meters. Having water on tap is transformative for the lives of hundreds of women and their families, it frees them from a time consuming and exhausting daily chore. It would have been worth installing the fog harvesting system even if climate change wasn't having an impact. Dar Si Hmad is a female led and local NGO, so understand this, it is why they didn't stop once the system was in place. That was never the intention, they look at development holistically.

As well as overseeing the installation, maintenance and expansion of the fog harvesting system, Dar Si Hmad work to further empower the rural Berber women in the project villages, thereby multiplying the benefits of the project. The women, who are keen to use the extra time they now have, are learning agro ecology techniques at Dar Si Hmad learning farm and attending workshops that help them to improve their digital and literacy skills. They are also being trained in how to manage and maintain the fog harvesting system and working together to develop income-generating projects. And success breeds success. Responding to interest from neighbouring villages, Dar Si Hmad have partnered with German company Aqualonis to upgrade to the more advanced CloudFisher model and now have 1,700m$^2$ of fog harvesting nets in place. The system now supplies water to eight more villages, with women there also participating in Dar Si Hmad's education programmes.

As more regions become water stressed due to climate breakdown, more fog harvesting projects are appearing. There are examples in other parts of Africa, but also in South and North America, Europe and Asia. The achievements of Dar Si Hmad are inspiring: fog harvesting has clearly been a success. It is tempting to dash off and purchase a CloudFisher for every fog covered hill farmer in the world. But it is important to remember why this Moroccan project is so celebrated. This requires us to recognise that it has been a technical *and* sociocultural success, with the emphasis on the latter. Ultimately, whilst the equipment and grid are impressive, it is not the technology itself that is being celebrated – anyone with a steady supply of fog can install a fog catcher – what is being celebrated is the transformative effect this choice of adaptation strategy is having on the people of Aït Baâmrane.

Fog harvesting has significantly improved lives in Aït Baâmrane by making the previously impossible, possible. It has therefore become a piece of infrastructure that the community values highly. This means that when external funding and grants are eventually withdrawn, the community will be motivated and ready to take on the responsibility for running and maintaining *their* fog harvesting system; it won't go to rust. Dar Si Hmad's holistic approach is vital to this and is the key learning here: technological innovations are only as good as the social, economic, environmental, and educational initiatives that are wrapped around them.

Fog harvesting has huge potential, but it will only succeed at scale if implementing bodies work to build the structures that will support it over the long term. Dar Si Hmad is not just a local NGO, they are *of* the community rather than merely being *in* it or *for* it. They are showing the way. Their model is replicable, but it has to be right for the community it lands in. The model must be the community's model, nobody else's.

## So you've declared a climate emergency?

Politics is often cited as the chief barrier to action on climate change. This seems especially true at the national and international level, where vested interests stymy progress, inertia reigns, and voluntary commitments (that are often inadequate) have become the norm. However, there are some encouraging signs that, at a local level, political willingness to take radical action is starting to grow.

In early 2020, less than a month before Boris Johnson announced that the UK was to go into a Coronavirus lockdown, a Full Council meeting was held by Warwick District Council (WDC) to vote on a proposal to increase council tax by an unprecedented 34.2%. The proposed increase, £57 per year – which would only be levied on the highest value properties – would create a ring-fenced fund to enable WDC to roll out a Climate Emergency Action Programme[83]. Councillors voted the proposal through, a move that briefly made the national headlines[84].

Warwickshire is in central England, south-east of Birmingham, north of Oxfordshire. It is middle-England, both culturally and geographically, and not the first place you'd think of as a hotbed for radical climate policy. WDC is one of five district councils in Warwickshire and covers an area that contains both the county town of Warwick and its near neighbour, Leamington Spa. It is home to approximately 150,000 people.

Warwick District Council is normally a stronghold for the Conservative Party but, in the May 2019 local elections, that changed. Support surged for the opposition parties, especially for the Green Party and the Liberal Democrats, who gained seven seats each. As a result, two distinct and evenly balanced groups have emerged. Between them, Labour (5 seats), the Liberal Democrats (9 seats), and the Green Party (8 seats) now hold 22 seats and form one group on the 44-seat council. The Conservatives fell from 31 to 19 seats and combine with Whitnash Residents Association (who held onto their 3 seats) to form the other group.

What is surprising, and encouraging, is that rather than descending into a politics of stalemate, the two groups have come together to vote unanimously in favour of both a declaration of a Climate Emergency (made in June 2019) and the proposal to increase council tax for the highest value properties. The political will for radical action on climate change appears to be there. However, the Council can't impose an increase in council tax – which equates to £1 extra per week per household – without specific support from their constituents. By law, the proposal must go to a public referendum. The politicians are behind it, but are the public?

If it wasn't for the Coronavirus pandemic, we would already know. The referendum had been scheduled for May 7th 2020, but it was postponed, potentially permanently. To fill the void, WDC organised a climate change 'People's Inquiry' which took place, online, either side of Christmas 2020. During the inquiry, a representative sample of 30 local citizens came together seven times[85]. They heard from a diverse range of climate change experts, discussed the evidence they were presented with, did their own research, and worked up a 'Jury Statement' and set of recommendations to put to the council.

The statement (printed in full on page 38) and recommendations made by the jury, will have encouraged the councillors. 27 out of 30 jury members voted on the recommendations. 21 *strongly supported* the statement, three *supported* it, two people voted *neither to support nor oppose*, and only one was *strongly opposed*. This high level of support suggests that the public will does exist, a referendum on a council tax hike would test it.

If, in the wake of the Coronavirus pandemic, the referendum does eventually go ahead, Councils of various sizes all over the country (and possibly the world) will be watching with interest. Warwick District Council could set a very powerful precedent and trigger a landslide of bottom-up, locally led action on the climate emergency. Even if the referendum is permanently cancelled or postponed, or if it goes ahead and residents vote against the rise, Warwick have still set a precedent that is hugely encouraging - and possibly a sign of things to come. If a council made up of politicians from across the political spectrum can unanimously vote in favour of a proposal to increase council tax by 34.2% for the wealthiest – just to take action on climate change – surely other councils can too.

So, what motivated Councillors of all persuasions and stripes to support such a radical proposal? It is unclear, exactly, but there are clues in the main report of the Climate Emergency Action Programme and the CAN (Climate Action Now) campaign[86] that launched as Councillors were voting on the council tax proposal. The CAN campaign has cross-party backing, is fronted by the Councillors themselves, and exists to persuade voters to vote 'yes' in any forthcoming referendum. CAN is heavily focused on efforts to tackle the causes of the climate emergency. It therefore directly responds to the growing awareness of the need to drastically cut greenhouse gas emissions (something the People's Inquiry recommendations echo). Councillors were extolling the win-win benefits of the emissions reduction plans they were proposing: greener streets, a decrease in air pollution, better public transport, more parks and recreation spaces, jobs in the green economy and so on.

The detail in the main report of the Climate Emergency Action Programme reveals more. While WDC are making it clear that by *'Taking Action on Climate Change'* it is possible to make *'Warwick District a great place to live, work and visit - and carbon neutral by 2030'*, they are also stressing the need for adaptation strategies and detailing how they plan to implement them. WDC has recognised the direct impact climate change will have on its residents and that, perhaps, central Government is not going to have the wherewithal to come to their rescue. They are especially concerned about increased frequency and intensity of floods and heatwaves and stress the need for: *'Forward planning rather than reacting to extreme weather events as they occur.'* They are keen to ensure that the Council itself is adapting, so that key pieces of infrastructure don't fail on them, but they plan to also encourage and guide businesses, institutions and community organisations to include adaptation strategies in their long-term business planning.

There are also promising signs that the adaptation measures promoted will complement efforts to cut emissions and improve biodiversity. They will, for example: *'support adaptation measures such as tree planting for carbon sequestration and 'cooling-off' benefits.'*[87] The thinking appears to be holistic and long-term.

Building on recommendations from the 2017 UK Climate Change Risk Assessment[88], WDC have identified four key action areas:-

1. Flood planning to cope with increased rainfall;
2. Health and wellbeing contingency planning to adapt to higher temperatures and heatwaves;
3. Planning for the protection of the public water supply;
4. Planning for the protection of natural capital.

These adaptation plans are receiving very little fanfare from Councillors and the CAN campaign, they are far more focused on the mitigation plans. Adaptation experts were also conspicuous in their absence at the People's Inquiry. WDC and the CAN have apparently calculated that it is concerns about climate chaos that will win support for climate action and – if it goes ahead – votes at the referendum.

This mitigation focus is repeated in most places where a climate emergency has been declared. The hope is that the mitigation efforts of the people of Warwick District will combine with action being taken elsewhere, and that the combined effort will slow the acceleration of climate breakdown, reducing its impacts. However, the type and scale of action taken elsewhere is out of Warwick's hands. Adaptation to climate breakdown, on the other hand, isn't. There is a lot that can and should be done at the local level and, if the yes vote succeeds in any future referendum, a lot will be done. WDC's leadership, foresight and long-term commitment to the wellbeing of its citizens is a very encouraging sign of hope. It is also an example for others to follow.

## It's a long way from Warwick to Deusa

The difference between life in prosperous middle England and life in the remote mountain villages of Nepal is difficult to quantify. Whilst inequalities exist in Warwick, it is still one of the wealthiest towns, in one of the wealthiest countries in the world. Deusa is in Solukhumbu, eastern Nepal. It is overlooked by Mera Peak mountain and approximately 60 km due south of the mighty Sagarmatha (Mount Everest). Deusa is categorically not wealthy. It is one of the poorest places in one of the poorest countries in the world. Yet both Warwick and Deusa will need to adapt to climate change.

Climate change impacts Nepal in multiple ways. The most eye-catching effect is high mountain glacier retreat, it is totemic. Elsewhere, however, in communities like Deusa, the effects are not as immediately obvious. They take the form of landslides, droughts, flooding, insect pests, water shortages and heatwaves. These all combine to make rural life very challenging, but adaptations are, thankfully, possible. Enabling them requires resources, training, expertise, commitment, openness to change and, of course, funding.

The Glacier Trust (TGT) was founded in 2008 by Robin Garton. He wanted to raise money here in the UK and grant it to organisations in Nepal who were enabling climate change adaptation and education. Sadly, I never met Robin. He died, tragically, while on a solo trek in Scotland in 2015. To this day, nobody knows exactly what happened to him. His disappearance, the long and uncertain search and eventual discovery of his body, was devastating for his family, friends and everyone associated with TGT. He is fondly remembered both here in the UK and in Nepal.

Since joining The Glacier Trust in late 2016, I have been able to get to know Robin's family and friends in both the UK and Nepal. I have shared meals with his widow, daughter, son and various nieces, nephews and cousins who are all great company and loyal supporters of our work. As I've learned more about Robin and his life, his influence on me has grown. What inspires me most about what he did is that he *wasn't* the chief architect and didn't want to be. As a funder, he made it clear that The Glacier Trust was very much open to persuasion. This inherent humbleness is at the core of TGT's work. We are not silent partners, we do chip in with ideas and insights where we can, but we are not the experts, we *enable* the experts - and they live in Nepal.

I am taking huge inspiration from Robin's deeply trusting approach and feel very fortunate to be witnessing the realisation of a vision that emerged as The Glacier Trust was establishing itself. Since those early days of TGT, that original vision has turned into a project, and is now rapidly becoming a movement. I am a bit biased, but what started in Deusa is a potentially *Great* Adaptation. I want to share that story with you to illustrate how a focus on agro forestry, agro ecology and agro biodiversity can intersect with objectives on gender equality, sustainable development, climate change mitigation *and* adaptation to usher in progress across multiple agendas.

Robin spent a lot of time in Nepal, met many specialists in climate change, development and community organising. He

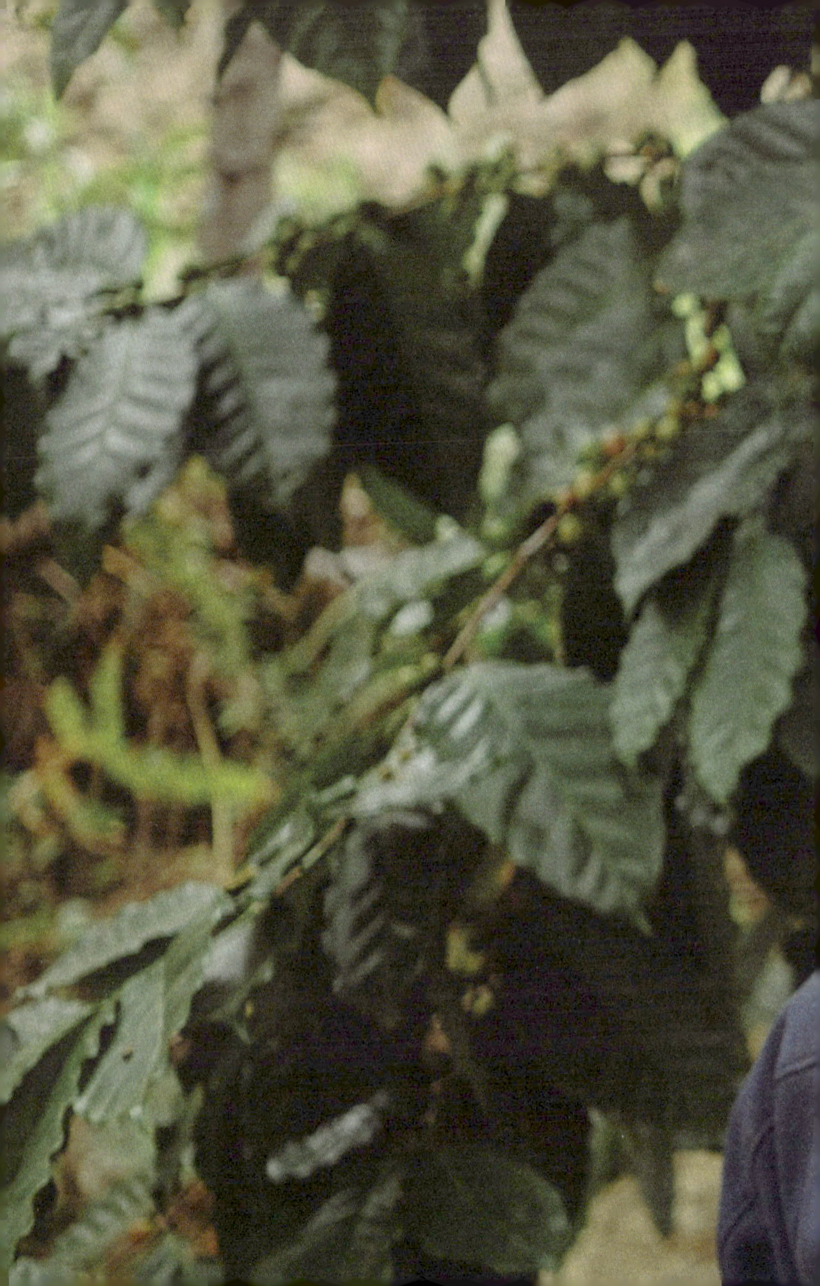

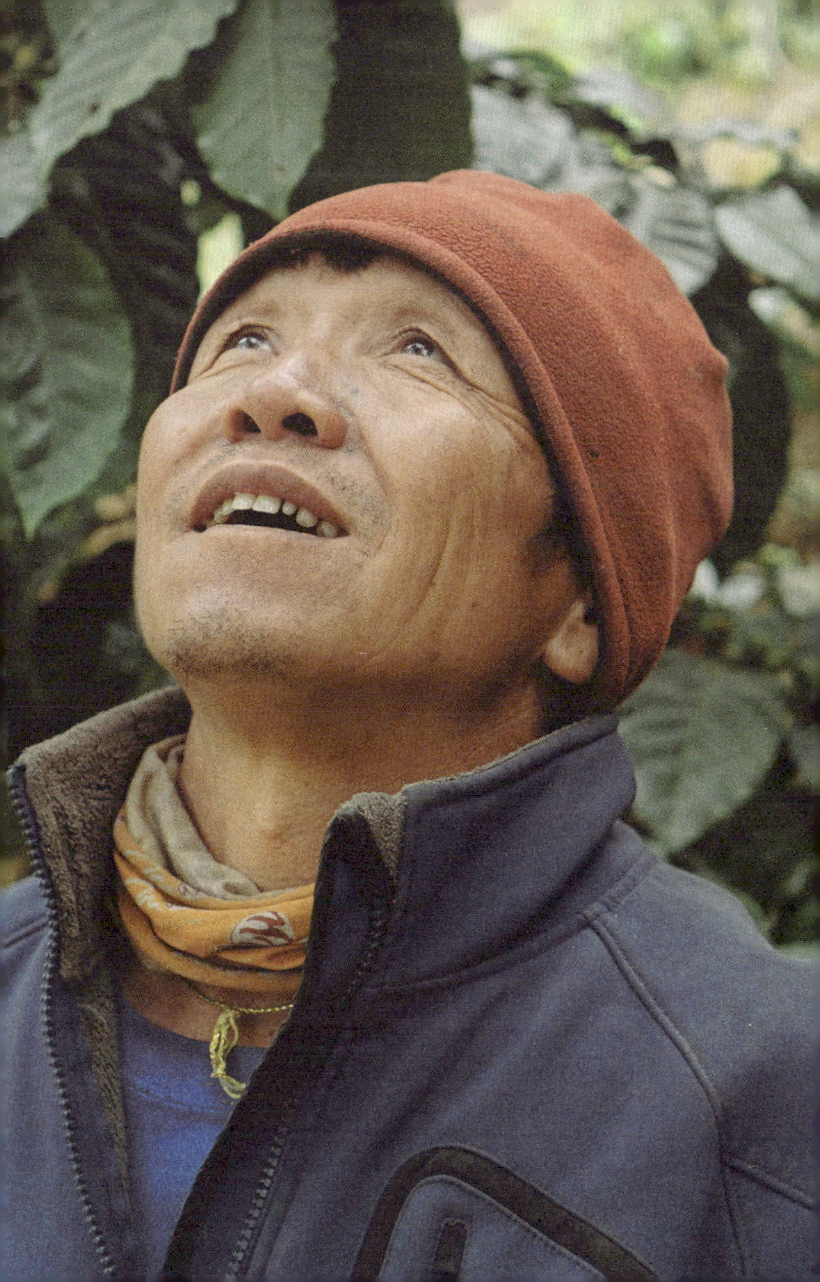

built up working relationships that were more like friendships and created bonds between TGT and partner NGOs that are still very strong today. One of the most important relationships Robin developed was with Narayan Dhakhal. Narayan is Executive Director of EcoHimal Nepal, an independent but globally networked NGO. He is a thinker, mobiliser and doer. He splits his time almost 50/50 between his office in Kathmandu and his true office, the mountain villages of Sankhuwashabha, Kavrepalanchok and Solukhumbu.

Whilst trekking together in Solukhumbu in 2007, Robin and Narayan (who had not long met) were discussing the potential of organic agro forestry as a strategy for development and climate change adaptation. Narayan was introducing Robin to the handful of farmers in the area who were adopting agro forestry and having success, both ecologically and economically. As they walked between villages, they were discussing what needed to be done to help spread the practice of agro forestry and agro ecology in these remote mountain villages. Narayan presented an idea to Robin that had come to him via farmers and community leaders in and around the sub-district of Lokhim. Their idea was to create a physical centre for agro forestry in the heart of the community. It would be run by the people, for the people, and enable climate change adaptation in ways that complement efforts to tackle the root causes of the other problems vulnerable and marginalised people face. The idea – a community-led Agro Forestry Resource Centre (AFRC) – was born.

At that time, Nepal was just emerging from a ten-year civil war, it was difficult to get new projects off the ground. The Glacier

Trust was in its very early stages too. Robin couldn't immediately offer EcoHimal the funding it would need to start the project. But he promised Narayan that if he could find a local community who wanted an AFRC he could count on The Glacier Trust to raise the start-up funds. On returning to the UK, Robin urged Narayan to develop the concept further. Narayan set out to present the AFRC idea to as many people as he could, right across Solukhumbu. He wanted to see how much enthusiasm there would be for it. He gathered ideas and insights and strove to find a way to make an AFRC happen.

In the late 2000's and early 2010's, the villages of southern Solukhumbu were in danger of heading in the same direction as so many other rural communities in Nepal that were suffering out-migration and a growing number of social and environmental problems. EcoHimal were already working on sanitation and education projects in these villages. They were working with the local community to build schools, sanitation and water supply systems. This work was giving children access to education, putting a permanent end to open defecation, and improving overall living conditions for thousands of families. But Narayan was concerned that the gains from these developments would be lost if the area didn't also develop socially, ecologically and economically.

With awareness of the impacts of climate change starting to increase, the danger that land and communities might go into rapid decline was starting to loom. The economic growth strategy therefore had to be climate smart and ecologically smart. It needed to be socially, economically and environmentally sustainable in a context that was fragile on all three fronts. Organic agro forestry, combined with community organising and cooperative business models, seemed to offer a way forward. This was core to the Agro Forestry Resource Centre vision that Narayan was starting to share with decision makers at all levels in Solukhumbu.

Through this sharing, a more detailed vision for a community-led Agro Forestry Resource Centre began to emerge. The perfect AFRC would have a central building that would act as a focal point and a hub for people to gravitate towards. It would house an office, a training hall, dining facilities, a kitchen and accommodation. It would also have a working example of a kitchen garden as part of a wider 'outdoor classroom' to *show*, not just *tell*, people about agro forestry, organic farming and climate change adaptation.

Narayan explained the vision to as many people as possible. He was met with scepticism over where it would be located, who would run it, how it would pay for itself and who would benefit. But as the concept became clearer in people's minds, enthusiasm for it grew. The community-led ethos that is core to the model guards against the 'elite capture' that so often dogs development projects. The AFRC wouldn't land in a village and make the rich richer, it wouldn't exacerbate existing inequalities and injustice. It would, if run well, do the opposite and the more marginalised and vulnerable people Narayan was talking to could see this. It would be easy for them to become members of the AFRC and they would play a role in designing and running it.

After two years of drumming up support, an offer came Narayan's way. He was speaking at a district council meeting in Salleri, the administrative capital of Solukhumbu. It was attended by over a hundred movers and shakers from the district - politicians, officials from various local government departments, NGO staff and leaders, the media, and interested citizens from right across southern Solu. Narayan's talk went well, people liked the idea, they understood the concept, the kick-starting role EcoHimal and The Glacier Trust would play, and the *for the people, by the people* ethos.

Finally, after many questions, answers and debate, the conversation died down. Narayan was starting to thank people for their time and ideas when the local councillor for Deusa, Ram Shangrarhsa Kirati, raised his hand. I have met Ram a few times now - he is a true representative of the people he serves, he understands their needs, aspirations, and spirit. He had raised his hand to make a pitch to Narayan. He wanted the AFRC to be built in Deusa and he was willing to put some of his local government budget and resources towards getting it started. This was the opportunity Robin and Narayan had hoped for. At a community meeting two months later, Ram and Narayan outlined the AFRC vision to the people of Deusa. It went down well, and an initial committee was formed. Soon after, secretary of the committee, a local lead farmer, Tilak Rai, invited Robin, Ram and Narayan to work with him on a plan to find a suitable site for Deusa AFRC. They agreed on two principles to guide their search. The AFRC should be:-

- near a road, so that resources and people could come and go with relative ease (relative because these are not metalled roads).

- on marginal land - i.e., land that wasn't currently being used for agriculture because it was thought too steep, too difficult.

This latter principle had two advantages. Firstly, it would be easier to persuade a landowner to part with marginal land and, secondly, if Deusa AFRC could prove that organic agro forestry and all the other things it wanted to demonstrate were possible on marginal land, it would give farmers confidence that it could also be successful in more hospitable locations.

Robin travelled to Deusa to team up with Tilak Rai to start the search for a site. They were joined by Mohan Rai and Sangita Shrestha, who were already working for EcoHimal in the area. Together, they trekked through Deusa searching for the right location. Eventually they found somewhere they thought would work. The land, just below a road and close to a small hamlet, was steep and not in use by anyone other than the occasional roaming goat. It was, however, owned by four different people. But what initially felt like a complication turned out to be an advantage. All four landowners – Mr. Ram Bahadur Rai, Mrs. Maina Kumari Rai, Mr. Harka Bahadur Tamang and Mr. Jeet Bahadur Rai – on hearing the plans, agreed to donate their quarter of the land to the project. They became founder members of Deusa AFRC's board and their universal approval of the project showed once again how popular the concept was with the community. Four more AFRC advocates had been recruited.

It took just over a year to set Deusa AFRC up as an independent NGO and organise the transfer of the land over to it and therefore the community. By the end of 2012, Deusa AFRC had land, a board of directors and the support of the local community. It also had a daunting task ahead of it. The land was stony, steep and overgrown. For the vision to become a reality, a lot of work was needed. What happened next confirmed exactly how much support there was in the community.

Motivated by Narayan, Ram, Robin and the dream of what the AFRC might make possible, 1,000 people from right across Deusa volunteered a total of 3,000 days labour to the task of preparing the land. Together they created a level platform big enough for the main centre building, a toilet block, a kitchen block and an office. Below and to the side of the main buildings they cut a ten-step terrace to create 250 hectares of farmland. The terraces were built to grow plant seedlings, tree saplings and organic vegetables. These are all now thriving, with a huge diversity of crops growing in the open and under polytunnels. Most of what is grown is for sale in the local community, which has increased the amount and variety of produce available in Deusa, improving diets and improving agrobiodiversity. Other parts of the AFRC's land are used to grow crops that the agricultural specialists employed by the centre want to test in a controlled environment. They want to establish what grows well in the changing conditions. If they can do this, they will be in a position to recommend appropriate, climate smart crops to AFRC members.

By 2013, with the terraces built, training events started. Farmers came from the local area to learn the best methods for planting fruit trees, how to create a kitchen garden, what could be done to protect crops from insect pests, and how to manage water effectively. This training has continued every week since and farmers of every gender are welcome. Slightly over 50% of attendees identify as male, but there is strong female representation, and the group is diverse and inclusive - this has been a key to its success. The group is learning from specialists, but also from each other, they are farming organically and building resilience to climate change through agrobiodiversity.

In 2014, the foundations for the AFRC's main building were dug, again by a local volunteer labour force. The three-storey building was completed in 2015, ironically just in time for Nepal to be hit by two devastating earthquakes. Fortunately, the building survived, suffering no structural damage. It was inspected and soon declared safe and ready for use. The ground floor is a small multipurpose hall, used for meetings, trainings, lectures and, on some mornings, as a children's nursery. On the top two floors there is a small office and three bedrooms, it is one of my favourite places in the world to stay the night.

Robin last visited Nepal and Deusa in late 2014. He was there to see the walls of the building going up, but sadly never got to see the finished building. A plaque dedicated to his memory has been installed at the main entrance, it is kept in immaculate condition, I read it every time I visit.

---

Elsewhere on the AFRC site, land was assigned to exhibit other practices that farmers could adopt. Climate change is impacting on the rainy and dry seasons in Nepal. The monsoon is becoming more erratic, steady predictable rain is a thing of the past. The dry season is getting even drier, so any rain that does fall is precious. At the AFRC, to demonstrate how water can be collected and stored, rainwater harvesting ponds were dug and linked, via drainpipes, to the roofs of the buildings. This simple adaptation to one of the impacts of climate change is easy to copy, especially if farmers are shown how to do it and given access to the tools for the job. This is one of the many things the AFRC does.

As the monsoon has become less predictable it has also become more violent. The trend is now for deluge and drought. The heavy downpours are one of the reasons that landslides are on the increase and even small ones cause problems: they damage buildings, block roads and paths, and can destroy crops on any plant beds down below. The footpath down to Deusa AFRC cuts across a very steep and rocky slope, which isn't suitable for terracing. However, instead of leaving it as scrubland, the AFRC team decided to plant 'broom' grass as a way to demonstrate what can be done to stabilise slopes to help prevent landslides. The broom grass has other uses too. The tops of the grass can be periodically lopped off to use as fodder for livestock. And, as its nickname suggests, it can be harvested and bound together into a broom. This technique is taught at the AFRC and farmers who produce a lot of broom grass can supplement their income by making and selling brooms in the local markets.

Deusa AFRC is also showcasing other mitigation and adaptation strategies. They have installed solar water heating on the roof of the bathroom block and encourage a wide variety of water saving techniques to apply on the farm and in the home. In addition, they are working hard to practice what they preach. They have composting toilets and use the end product to teach farmers how to make and use organic crop fertiliser. All of these innovations are shifting attitudes and perceptions of what's possible. Deusa AFRC and its users are a glowing example of how it is possible to thrive in the face of climate disruption.

---

So, what of the Agro Forestry? Does the AFRC do what it says on the tin? It is fair to say that Deusa AFRC is transforming agricultural practice in this corner of Solukhumbu. More and more farmers are turning to agro forestry as a climate change adaptation and sustainable development strategy. In essence, agro forestry is the farming of trees. They are grown, coppiced, and harvested to provide fruit, nuts, fodder and timber. It can be done in horribly mono cultural and chemical ways, but also organically and with a strong emphasis on agrobiodiversity. It is a nature-based solution to some of the problems climate change causes – landslides, heatwaves, food shortages – and a way for families to work the land in less labour intensive ways as economic forces pull the rural labour force away from remote mountain life.

In Deusa, thousands of farmers are being enabled to grow fruits, nuts and high-altitude speciality coffee. These are not necessarily easy things to grow, process and sell, but the training and tools being provided by EcoHimal and Deusa AFRC are making it possible. Typically, coffee is grown under the shade of banana trees, with root vegetables being grown at ground level - this is the 'layer farming' method and it is proving popular and profitable. An added bonus from growing trees is that they provide shade, and as temperatures rise that makes a difference, especially the intense heat of summer arrives. The newly forested areas are microclimates that have a cooling effect on people and animals. Trees also absorb a lot more carbon dioxide than plants like rice and millet, so there is a mitigation benefit accruing too.

As important as the training and tools are, the third ingredient is the most important - collaboration. EcoHimal have been integral to fostering this, they have worked closely with the AFRC and the local community to establish a fruit and coffee growers cooperative in Deusa. It has a gender balanced governing board and meets at the AFRC. Farmers there are working together to harvest coffee, process it to parchment stage and prepare it for transport to market in faraway Kathmandu. EcoHimal have helped to forge these links, finding buyers and roasters who give a fair price. The Glacier Trust has even managed to bring a few kilograms of roasted coffee back to the UK for sale as 'Nepal Glacier Coffee' to our supporters.

As well as coffee, Deusa AFRC is also trialling hazelnuts, macadamia and almonds. They hope to establish these trees in Solukhumbu to create more agrobiodiversity and, with it, resilience. This strategy also suits the aging population in rural areas. Tending and harvesting trees is hard work, but it is nowhere near as back breaking as the arable farming that is slowly disappearing as the young, fit, labour force required to farm it migrates to the city or abroad.

---

Between 2014 and 2017, demand for Deusa AFRC's services grew, and the need for an outreach strategy became clear. The solution was the formation of 'satellite plant nurseries'

प्रश्नविर
शाइ

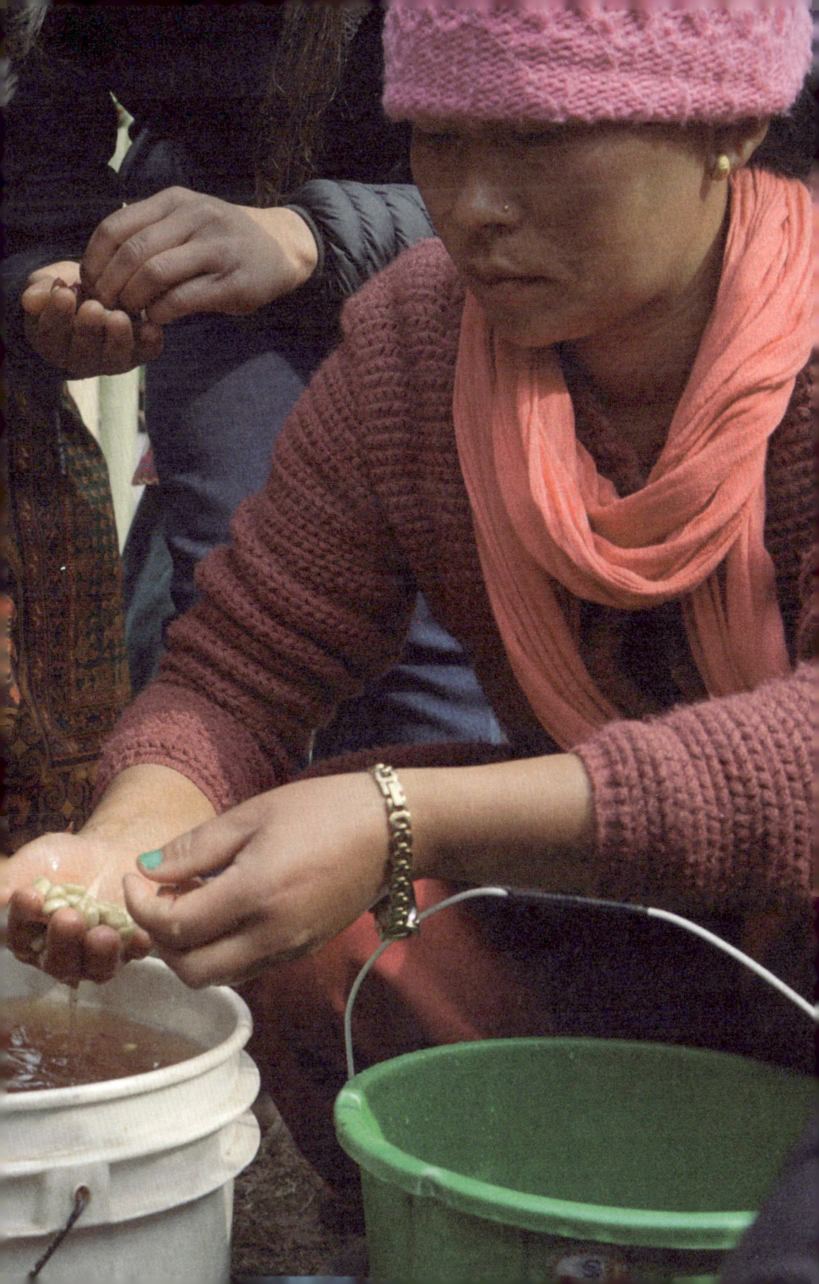

that act as *mini*-AFRCs. These satellites are run by experienced and skilled farmers on land they own and dedicate to this purpose. The satellites become hyperlocal centres for production and sale of seeds, seedlings and produce. They also provide new locations for the central AFRC team to deliver trainings and trial innovations. There are now seven satellite nurseries in Deusa, strategically located to serve as many farmers as possible, but they are also well spread geographically, which allows the AFRC's agricultural specialists to experiment with crops at different altitudes and microclimates.

This 'hub and spoke' AFRC model has gained attention and is now starting to be replicated elsewhere in Nepal. In 2018, TGT and EcoHimal were awarded a three-year grant to replicate the model in Mandan Deupur, Kavrepalanchok. We have since been awarded another three-year grant to extend the Mandan Deupur AFRC project through to 2025. EcoHimal have also been awarded a major grant to build three more AFRCs in the district of Khotang, just south of Solukhumbu, and another in Sindhupalchok. The original AFRC idea has gone from a vision, to a project, and now to a movement. This is climate change adaptation combined with sustainable development and we believe it can be replicated right across the Himalayas over the coming decades, potentially around the world.

## *Mi*grate adaptations

You could air condition every pavement, try a hundred different varieties of grape, spray a whole mountain in fake snow, set up a cool room on every street corner, and catch all the fog you could possibly ever need, but – to adapt well to climate change – you don't necessarily have to. Sometimes you just move. Climate migration is so often framed as a negative thing, an option of last resort that is forced upon people who would far rather stay put. In the case of a flooded Pacific Island that can be the case, but there is a lot more nuance to the story of climate migration.

---

Right across the global South, there are rural areas that are suffering from environmental stress. Climate change is one of the biggest drivers of these stresses, particularly when it impacts on the supply of water to farms and farmers. For a growing number of households, migration is one adaptation strategy amongst many and it is often the one that makes other adaptation strategies possible. The key thing to hold in mind here is that migration can be seasonal, temporary and cyclical. People are not always talking about permanent migration, where a whole family abandons home in search of new life. They can be talking about one or two members of a household migrating for short periods to another location, with an expressed intention to return - or at least come and go. For example, they might migrate to find work, develop new skills and knowledge, make contacts, and to earn money. This enhances their adaptive capacity, allowing them to invest in the equipment and resources needed to make adaptation to climate change possible at home.

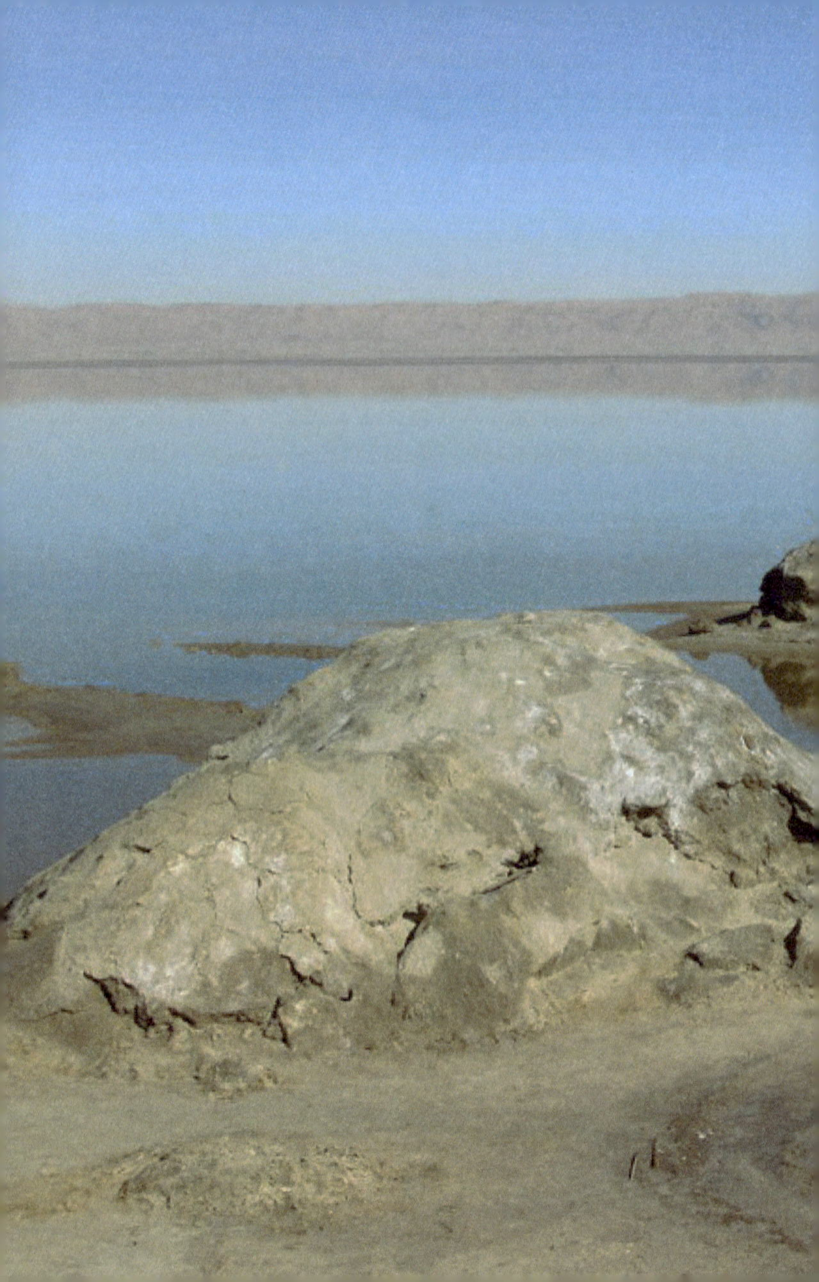

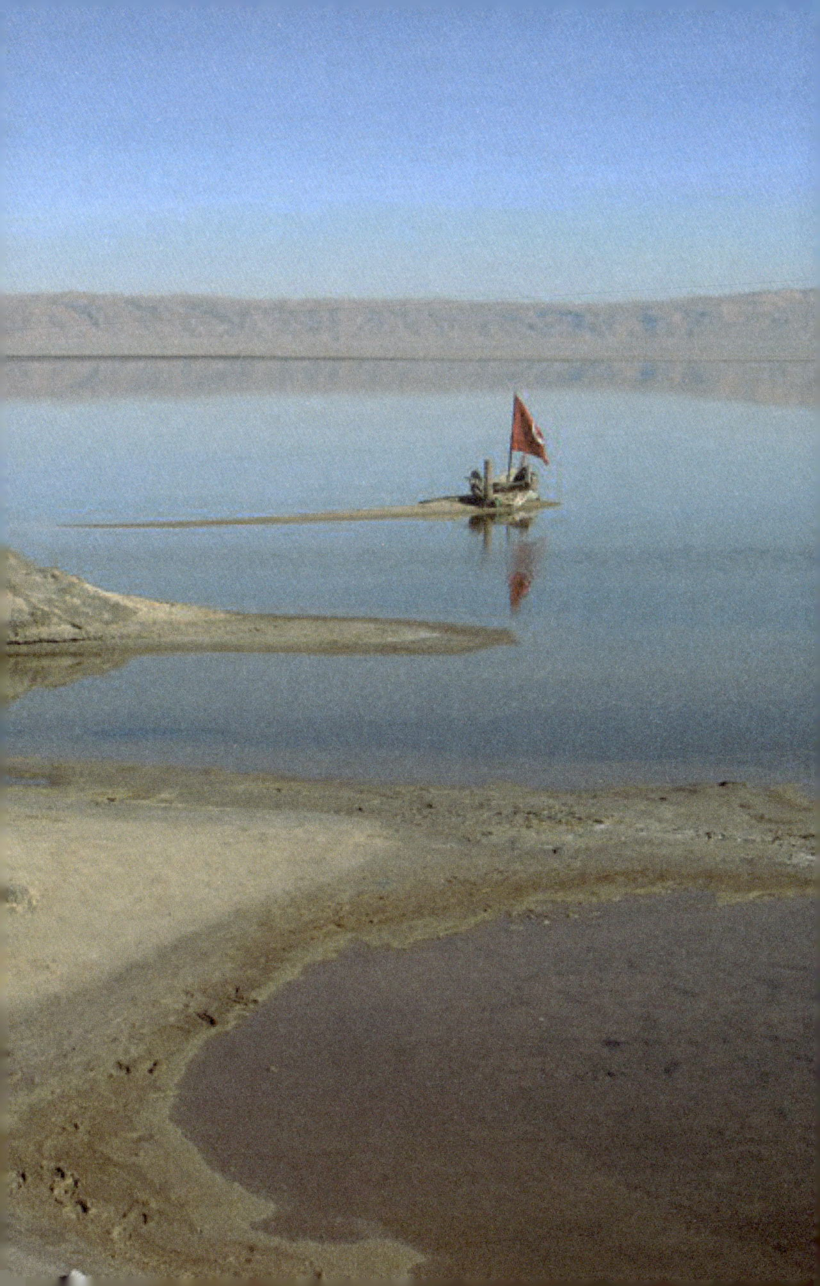

In El Faouar, central Tunisia, just south of Chott el-Jerid, a seasonal saltwater lake, climate change is exacerbating an already water stressed situation. Households there need to adapt to survive, especially those who rely on agriculture. Two researchers, Karolina Sobczak-Szelc and Naima Fekih, have studied how families seek to diversify their income by labouring, working in tourism, and how they cooperate with neighbours to share the costs of buying the equipment they need to drill for wells for water[89]. Seasonal migration to other rural areas, as well as to towns and cities, and abroad, is a big part of this attempt to bring in new income to invest at home and shore up a precarious situation. Migration, for many El Faouar households, isn't turned to as an adaptation of last resort, it is a proactive decision taken to stave off forced migration of the entire family in the future. One interviewee told the research team about their experience:

> *I combine sources of income; I sell dates, work in France, I work with a tractor because I have it for one month, one and a half, then it goes to the neighbour. This tractor, and some money from France I mix. ( ... ) Now, I build a garage, I prepare a place for selling agricultural materials and agricultural products. ( ... ) You have to mix the sources of income. You have to do this way in order to live here. (Interviewee H121, male, Sabria).*

Migration as adaptation is a positive thing, it enables people to remain (if only seasonally) in the place they call home, rather than having to abandon it totally. Migration is also, of course, not a new phenomenon. Humans are nomadic creatures - in the long history of our species the tendency to settle in one place is a relatively recent development. Unfortunately, however, a *'sedentary bias'*, that frames migration as a negative, has taken a strong hold. This thinking has permeated much practice in the fields of development and climate change adaptation. The default objectives of many well-meaning projects and programmes centre on doing what is necessary to prevent people from having to migrate, rather than viewing migration as an adaptation strategy that can make other adaptation strategies possible. Of course, there is a tangled web of complications associated with promoting *'migration as adaptation'*, it may be the last thing a household wants to do. However, if policymakers were to relinquish their sedentary biases, they could develop ways to enable migration as adaptation for those who would like to do it but do not have the means or opportunities to do it autonomously.

The successful examples of migration as adaptation that organisations such as the Climate and Migration Coalition[90] are highlighting may help us to see migration as a positive enabler of adaptation. Migration can be part of the adaptation mix, it doesn't have to be permanent, and it definitely doesn't have to be a negative step. In fact, migration in its cyclical, seasonal, forms can delay forced migration for a generation, possibly two, possibly forever. If a household wants to erect a fog catcher, why should they wait for a grant or a loan to pay for it? Why not enable household members to migrate to locations that need seasonal labour so that they can earn money to send home to buy a fog catcher outright?

# PART 3 – Transformation

> *The problem with populist climate movements, such as Extinction Rebellion and FridaysForFuture, that have fueled the recent declarations of climate emergency is, ironically, that they are not thinking big enough. Imposing the discipline of an emergency on the politics of climate change narrows the policy gaze to the restrictive logic of equating human well-being with reduced carbon emissions, implying in essence that the world would be a better place with fewer carbon emissions.*
>
> *What is needed now is to look at the world through a much broader lens than that provided by the singular policy goal of securing net-zero carbon emissions by a certain date—what I call 'hitting the carbon numbers.' The world continues to change at a dizzying pace and*

*is facing significant challenges in the shape of endemic geopolitical conflicts, shifts in economic and political power, and resurgent political nationalisms. A fourth industrial revolution is also under way, with rapidly emerging technologies in the fields of artificial intelligence, genomics, materials science, and digital communication, all with vast potential to change how people of the future will live, work, and govern. Societies are also experiencing increasing economic inequality, fragmentation of social trust, and new forms of skepticism about scientific knowledge.*

*Arresting climate change, whether in 10, 50, or 100 years, will have to take all this into account. Failing to do so could make the world a worse place even if powered by zero-carbon energy.*

Mike Hulme, 2019[91]

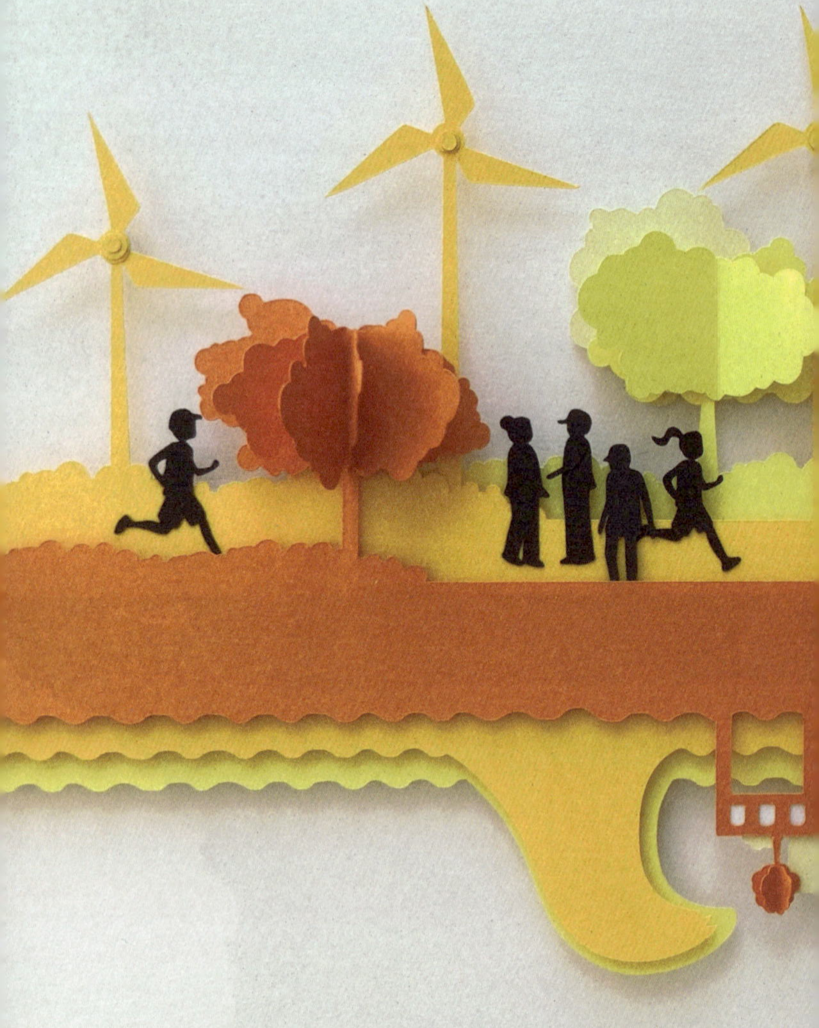

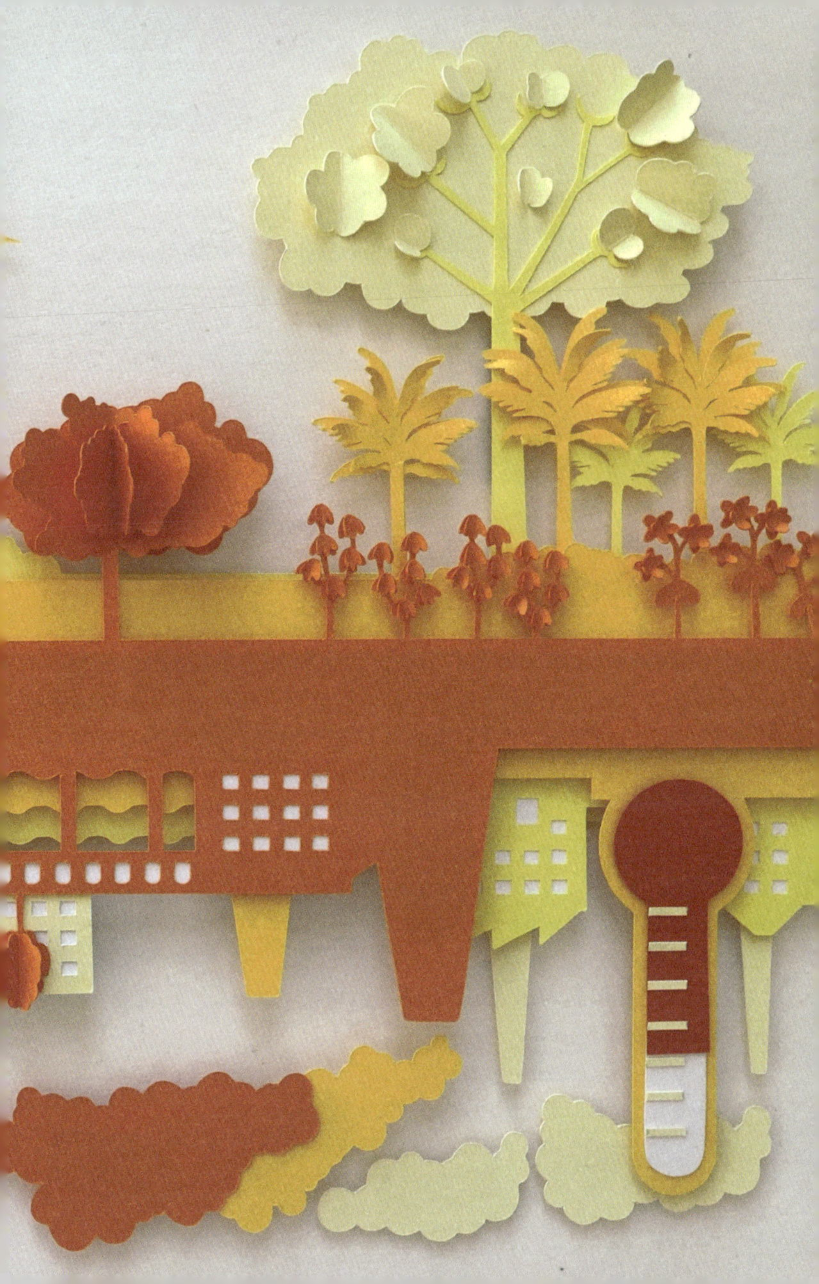

## 8. 'It looks bleak. Big deal, it looks bleak.'

So far, this book has covered adaptation from 28 different angles, but there are many more. A second volume could cover the adaptations being made by architects, insurers, sporting institutions, supermarkets, hospitals, energy companies; the list goes on and on.

Some adaptations are great, others are less so. Some make you smile, others make your toes curl. In these pages there are, I hope, plenty of talking points. Providing them is this book's primary purpose. There are, however, wider talking points that can't be ignored. Much of what you have read so far falls into the 'incremental' category, they are adaptations designed under an assumption that climate change will progress slowly and incrementally, while everything else remains broadly the same. But it might not. It is time to do what Mike Hulme implores us to do: *'look at the world through a much broader lens'* [91].

When it comes to climate change, it is still very difficult to say with any precision how bad things will get, or when, or where. The climate is a complex thing to understand and predict - it could be more, or less, sensitive to the emissions it is being prodded with, nobody truly knows, at least not definitively. And, for all the talk of this being a #RaceToZero, there is great uncertainty about how much action will actually be taken to reduce emissions - there is, in fact, a lot less certainty about this. The planet will warm, that is known for certain, but we don't know by how many degrees. So, precisely *how much* change needs to be adapted to is unclear.

These forecasting problems arise because the climate system – complicated enough to understand on its own – is an intricate system that interacts with a multitude of other complex systems and sub-systems. Working out how these systems bend and twist each other is not easy, not even in the present moment. What they will be doing to each other over the next twenty or thirty years feels almost impossible to work out.

Thankfully, not everyone is put off by all this complexity. Scientists wrestle with it and try like crazy to model every possible scenario. This results in a wide range of visions about the future that vary from the wholly terrifying to the comparatively hopeful. The mind boggles when it thinks about climate change, artificial intelligence, geoengineering, digital technology, biodiversity loss, space travel, democracy, religion, genetics, the whims of billionaires, and how they are all going to influence each other over the coming decades. The future is a mysterious place, it is as easy to land on a cheerful outlook as it is to end up in full blown despair. But the wider talking points we must confront are these:-

- What if things *do* get really bad?
- What if the climate system rises up and starts to bully all the other systems?
- Will adaptation – as currently done – still be possible?
- Will 'Western' civilisation survive, metamorphize, or completely fall apart?

In the rest of this book, 'civilisation' is a shorthand for the dominant, collectivised ways of thinking that influence and structure social, economic and cultural patterns and attitudes of human interaction with the planet we live on. One form of civilisation, crudely but aptly summarised as 'Western' civilisation, has become destructively hegemonic. Overpowering.

It is characterised by a neoliberal capitalism that privileges a few at the expense of the many. It creates vast inequalities both within and between countries: it is obsessed with economic growth, steamrollers (or commodifies) ancient traditions, globalises trade and culture, and treats humans, animals, plants and minerals as resources to be exploited. It pollutes, colonises, discriminates, destroys; and it riddles billions of people with chronic and acute mental and physical illness. Looked at through a macro lens it really isn't all that *civil* at all.

However, because 'Western' civilisation provides comfort to a few – a big enough and powerful enough few – it survives and spreads. **Some live comfortably *within* it, many more live uncomfortably *with* it.** And beyond them, there are only a tiny number of indigenous and peasant societies who still live truly outside of 'Western' civilisation. Their preciousness is rare, but they aren't necessarily something to seek a return to.

There are a growing number of people in the climate movement today who have reached the conclusion that things *are* going to get really bad. They are advancing the idea that 'Western' civilisation is in a downward spiral of rapid and profound self-destruction and may not survive. Part 3 of this book gives this idea serious consideration and examines how it is transforming notions of adaptation and the future.

## Two degrees of separation

In January 2019, The Glacier Trust wrote to the Royal National Lifeboat Institution (RNLI) to ask if they had a climate change adaptation strategy. After a phone call, several emails and a trawl through their website and reports, it was clear that if they did have one, it was light touch at best. We found that although there was a clear corporate line on the RNLI's commitment to tackling the causes of climate change, the impact climate change might have on the RNLI (and the people it exists to serve) did not seem to be a large factor in their decision making.

This has now thankfully changed (see chapter 11), but at the time little thought seemed to have been given to the impact that increases in the frequency and intensity of extreme weather events might have on the demand for the RNLI's inshore, offshore and inland services. The effect of sea level rise on the development and maintenance of Lifeboat stations and other elements of its coastal infrastructure also seemed to be largely overlooked. We were left wondering why climate change adaptation was not on the agenda for the RNLI? We wondered whether the RNLI were a one-off example, a rare case of a national organisation who had taken their hands off the tiller, or whether they were no different to the many other social actors in being so apparently complacent? It is likely they aren't, they join many others who have bought into a story about climate change that the UK government has been telling itself and others for decades.

The UK Government's story on climate change is a familiar and reassuring one (chapter 11 explores it in more detail). It goes like this: high-level, government-led action is going to keep global average temperature rise this century to *well below* 2°C above pre-industrial levels and, as far as possible, pursue efforts to limit the increase to just 1.5°C. Ministers and officials then go on to talk about being a world leader in a global effort, guided by the best scientists, to achieve Net Zero emissions by 2050. Achievement of this target, so the story goes, is more than sufficient to keep the climate stable and business-as-usual viable - panic over. By failing to add to this story, which would be to lay out a detailed plan of the near-term climate action, the Government also communicates something else: action can be delayed, no need to do anything too radical on mitigation or adaptation just yet. Hence the lack of near-term planning by the RNLI and others, they are just following the Government's lead.

---

The *'Net Zero by 2050'* target[92] was set in response to the spring 2019 Extinction Rebellion protests and supported by a feasibility study[93] from the Climate Change Committee (an independent statutory body). Moments like its announcement come along every few years. They are usually generated by campaigners driving climate change far enough up the agenda to provoke the establishment into a response. The response is a pledge that is just adequate enough to mollify protestors, but not so bold that it upsets key voters, or the right-wing press. Its effect is to bring an end to that particular chapter of the story.

It is at this point that most of the media and therefore general public, NGOs, civil servants, business owners and – troublingly – quite a few environmentalists, stop paying attention. They tune out from climate change for a bit, scroll down to whatever comes up next on their screen and move on. I know, for one, I'm guilty of that.

Meanwhile, the more grown-up parts of the Government machine are off doing something quite different. They know that the true climate change story is quite different, it is not nearly as hopeful and reassuring as it is so often presented. In short, the Government knows that the 1.5°C threshold will almost certainly be passed, and quite possibly the 2°C, 3°C and 4°C thresholds too. They know this because it is clear to anyone who looks closely at the carbon reduction commitments made in relation to the 2015 Paris Agreement. What they reveal is that even if the UK were to achieve its commitments, it is very uncertain whether other countries will too, the vast majority of commitments made are non-binding. Furthermore, as the UN itself acknowledges:-

> *Even if countries meet commitments made under the 2015 Paris Agreement, the world is heading for a 3.2 degrees Celsius global temperature rise over pre-industrial levels* [94].

The reality is that any decarbonising feats achieved by an increasingly marginalised global player like the UK will be overwhelmed by the greenhouse gases still spewing out from its friends and foes in the near and far reaches of the global economy. The latest round of pledges, the Nationally Determined Contributions (NDCs) that are being submitted in the run up to COP26, are doing little to alter this reality.

Climate scientists who follow the politics, largely agree. As far back as 2017, they lined up to tell Andrew Simms[95] that keeping temperature rise to below 2°C is *'unlikely'* [Prof Andrew Watkinson]; *'not very likely at all'* [Prof John Shepherd]; that there was *'very little chance'* [Prof Stuart Haszeldine]; that it was *'on the fanciful edge of plausible'* [Prof Piers Forster]; that *'we have emitted too much already'* [Prof Glen Peters]; that there *'is not a cat in hell's chance'* [Prof Bill McGuire] and that there is *'no chance whatsoever at current level of carbon emissions'* [Prof Joanna Haigh].

Since 2017, and not until after the 2020 election victory of President Joe Biden, the most optimistic statement made about future warming has been this:

> *The recent wave of net zero targets has put the Paris Agreement's 1.5°C within striking distance. The Climate Action Tracker (CAT) has calculated that global warming by 2100 could be as low as 2.1°C as a result of all the net zero pledges announced as of November 2020*[96].

Sadly, the infographic associated with this statement shows that CAT have also calculated that current policies could push global warming as high as 3.9°C by 2100. So, although limiting warming to 2°C might look *marginally* more possible now than it did in 2017, it is by no means the most likely outcome. Indeed, there are plenty of scientists who are prepared to say that 2°C now looks impossible[97]. So, although nothing should be ruled out, it is probably not a question of *if* we will hit 2°C, it is a question of *when*, and then whether temperatures will go higher - and then higher still?

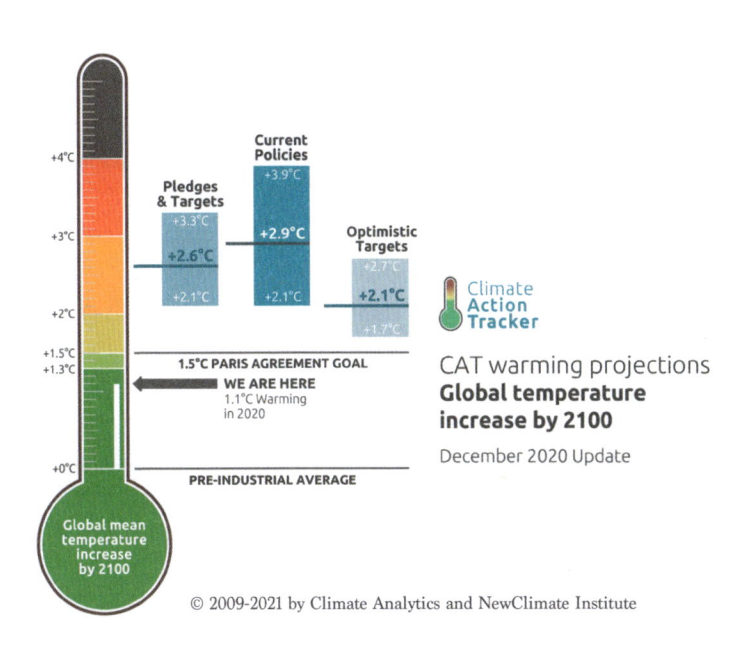

**CAT warming projections**
**Global temperature increase by 2100**

December 2020 Update

© 2009-2021 by Climate Analytics and NewClimate Institute

This book has so far been written with an implied assumption that if 2°C is hit, it will be the high point that precedes a stabilisation and, hopefully, a slow tumble back down to something resembling a 'natural' state. This scenario has been presented as a 'best-case' - with 'best' placed firmly within inverted commas. It is vital to emphasise that 'best-case' is a terrible choice of words: the *getting worse, then getting better,* scenario is nothing more than a 'least bad case'. The 1.2°C of warming that has already happened has wiped out homes, destroyed crops, triggered wars, shrunken glaciers, flooded entire communities and demolished ice sheets; the changes have been violent and rapid. In response, adaptation in all its planned, improvised and maligned forms has started. It will continue in all these ways and more over the next few decades as the atmosphere heats up.

If the final result of efforts to mitigate climate change is a temperature increase of 2°C, this is not something today's adults should be proud of. A 2°C rise could have been prevented, it should have been. 2°C will be fatal for so many people, plants, animals and landscapes; it isn't acceptable. It would also not - as the view through Hulme's *broader lens* reminds us - be an 'achievement' that guarantees a better world; climate change isn't the only game in town. Whatever happens now, the distressing truth is that it will not be possible to save everything we cherish. The losses will grow in number and magnitude, they will be painful and unjust. The mitigation and adaptation choices made will affect how much perishes and how much survives.

It is not that controversial to look at the science and politics of climate change and conclude that a rise in temperature of between 1.5°C and 2°C now appears to be the likely 'best-case' scenario. Many people in the climate movement are striving for this outcome, believing it to be the most rational goal to aim for. Either side of them are:-

- Those who insist that 1.5°C is still possible. They stick resolutely to their belief in 1.5°C and emphasise the moral and motivational imperatives of pursuing that target.
- Those who insist that the game is up, that the combined forces of economics, politics, and the greenhouse gases already in the atmosphere, make a future of catastrophic climate change unavoidable.

Let's call these two positions 'optimist' and 'pessimist', between them is 'not sure'. Arguments can be made for taking each of these positions. At this particular moment in history, it is possible to adopt any one of them and be granted a platform in mainstream debates. This is why splits appear in the environmental movement. I am in the 'not sure' position, but closer to the 'pessimist' end of it. This means that I believe 1.5°C to be extremely difficult to achieve, and 2°C to be slightly more possible but still very challenging. I also struggle to believe that 'Western' civilisation (as defined on page 142) can keep temperature rises within safe limits - consumerism and the relentless pursuit of economic growth look incompatible with a future of climate stability, healthy ecosystems and social justice.

My feeling is that, to have any hope of avoiding catastrophic climate change, 'Western' civilisation needs to be disassembled with great urgency and great care. I think this can be done but am very uncertain whether it will be. But if it were to happen, something new and better could be built in its place, something that is more ecological, just and climate safe. Anything other than deep systemic change is unlikely to be enough.

## Transformation, collapse, or total collapse?

Given how precarious the *getting worse, then getting better* 'best case' scenario now is, it is important to peer into the abyss of the 'pessimist' category. There is a real chance that climate change will fly straight through the 2°C threshold and the 2.5°C and 3°C thresholds and into a *getting worse, then getting – catastrophically – even worse* 'worst case' scenario.

Taking this almost apocalyptic stance reframes the debate entirely. It forces those who adopt it to confront existential questions of themselves and others. It can be a very disorientating and painful position to arrive at - and it forces a choice. The pessimist can either choose to retreat, hunker down, and look after number one in a bunker of self-imposed isolation - the path of the much-maligned *doomer*, who has given up. Or, they can get to work on transforming the world so that it can have the social and economic structures that truly *are* compatible with the adaptation, mitigation and social justice goals being set.

It is this second path that is of interest here. People who have started to follow it talk of how this stance frees them from the angst of fighting energy sapping battles to achieve carbon reduction gains within a paradigm that will only ever allow those gains to be marginal. They talk about the emergence of something positive and hopeful, making the case for creating it *before* collapse happens. So, whilst they are not optimistic about our chances of staying below 1.5°C, 2°C, or even 4°C, they are not nihilists, and they have not given up - they are still striving for deep mitigative action through systemic change.

There is nonetheless a danger attached to taking a downbeat stance on how successful mitigation efforts will be. The danger is that collapse preparation and adaptation might become too dominant a focus and completely crowd mitigation out. Talk of collapse also has the potential to legitimise those who have the appetite and resources to take a third path, one that involves unilaterally imposing large scale geoengineering experiments on the world - the *'climate behemoth'* future that Wainwright and Mann warn about in 'Climate Leviathan'[98].

It is unfair to accuse everyone in the pessimist category of wanting to take the first or third path. It is only a small minority who are abandoning all hope of mitigation, and only a tiny number of them are seeking to build themselves a lifeboat or a luxury survival bunker[99]. An even smaller minority (albeit a very powerful one) are seriously planning planetary-wide geoengineering solutions. So, we can rest assured, the doomists and the mavericks are still quite fringe. That is not to say that this won't change in the coming decades, widespread abandonment of mitigation is a troubling possibility - especially if the idea that planet Earth has entered a period of *catastrophic climate collapse* takes hold. And, be warned, lifeboaters *do* exist. In some cases, they aren't even being metaphorical: 'Seasteading'[100] is a thing - there are people out there who are talking seriously about building floating nation states at sea! As for unilaterally imposing large scale geoengineering projects, Elon Musk exists too[101]. Nobody knows what he, or Jeff Bezos, or any future Billionaire maverick, might do; they probably don't know it yet themselves.

---

The truly 'great' adaptations of the future will look at the world through the broader lens. They will be adaptations to not just climate change, but to the changes being witnessed in the rest of the natural world, in the world of technology, and in the arena of politics. If they are smart, they will be *malleable* (like many nature-based solutions are), rather than *hard* (like a concrete dam) - they will be adaptations that are themselves adaptable[102].

If things do get really bad, the 'great' adaptations of the future might also draw on the thinking behind two new concepts of adaptation: 'Deep Adaptation' and 'Transformative Adaptation' These two rapidly developing ideas and practices represent a step change in the adaptation discourse. They are steadfastly not 'incremental' in their approach to adaptation. They are, instead, approaches that search for – and respond to – paradigm shifts in how 'civilisation' is understood.

For anyone seeking to understand and advocate for adaptation, it is important to appreciate what adaptation actually means to those who want to go 'deep' and think 'transformatively'. You might not share their views on how bad the climate and ecological crisis will get, but they can still be your allies, they have ideas worth exploring. The sections that follow introduce some of the key thinking that is shaping the Deep and Transformative adaptation movements.

Two British academics – Rupert Read and Jem Bendell – have emerged as keynote voices on adaptation, their contributions have had a re-framing effect which has gone down better with some than others. Read was, until recently, a prominent voice in the Extinction Rebellion movement in the UK. Bendell is the founder and leading proponent of the provocative Deep Adaptation agenda. They are not adversaries, and agree on much[103]. But they do have slightly different takes on where the climate and ecological crisis is headed. Neither take is universally popular, but both provide a launchpad for a reconnoitre into what happens if things don't get better after they get worse.

Bendell's 'Deep Adaptation' exploded onto the climate change scene in the summer of 2018. It arrived in the form of a self-published PDF and was promptly downloaded and read by hundreds of thousands of people. It has had a mixed reception. Bendell has since amended the original paper and released a second edition in July 2020[104]. The revised edition made a number of clarifications and corrections, but the core argument remained. To summarise bluntly, Bendell predicts *'near term'* and *'inevitable'* societal collapse due to climate change. He bases this prediction on his own interpretations of the science and his experience as a corporate sustainability consultant. Bendell is calling for those who share his conclusion to consider its implications for *'research, organisational practice, personal development and public policy.'* His paper kicks-off an exploration of these implications and is an invitation for further debate and deliberation. To aid that process, Bendell offers up a 'four R's' framework: Resilience, Relinquishment, Restoration and Reconciliation.

Deep Adaptation is not in opposition to conventional forms of adaptation, but it has differences. Bendell's definition of 'resilience' helps explain how. He takes care to distinguish his use of the term resilience from the way it is typically used in the sustainability movement. For Deep Adaptation, resilience is *not* about developing a capacity to fully 'bounce back' from climate change induced shocks to resume a previous 'normal'. The resilience Bendell describes is more attuned to a psychologist's definition. He highlights how *'the concept of resilience in psychology does not assume that people return to how they were before.'* So, after a traumatic experience, psychological resilience shows up in a person's capacity to creatively reinterpret their identity and priorities. They recognise that their world, and therefore they – themselves – will never be the same again. They can still 'bounce back' but they will land somewhere different. The world will likely bounce back like this from the Coronavirus pandemic. Bendell urges us to embrace this 'less progressivist framing of resilience' when considering how to cope with the societal collapses he is predicting. The other three R's flow from this and talk about the need to:-

> *'let go of certain assets, behaviours and beliefs where retaining them could make matters worse'*
> [Relinquishment];
> *'rediscover attitudes and approaches to life and organisation that our hydrocarbon-fuelled civilisation eroded'*

[Restoration]; and
*'reconcile with each other and with the predicament
we must now live with'*
[Reconciliation].

The Deep Adaptation agenda has gathered momentum since 2018, interest groups have grown up around it on and offline. Bendell's ideas have also influenced some of Extinction Rebellion's leaders and, despite experiencing a degree of backlash[105] from within the climate movement, it looks set to inspire and infuriate in equal measure for a while yet.

---

Rupert Read is slightly more optimistic, and more moderate[106]. He argues that the current global hegemonic 'Western' civilisation is about to finish, but that this might result in social transformation, rather than social collapse. His vision of Transformative Adaptation (which he has dubbed *'TrAd'* to distinguish it from other visions of Transformative Adaptation) is more hopeful and less jarring, but many of the Deep Adaptation principles still apply; he does not argue against the need for resilience, relinquishment, restoration and reconciliation.

Read is not usually afraid of the limelight but predicting the end of civilisation is heavy stuff. He was nervous about the stir his theory might cause, so decided to lay it out first in an anonymous article titled *'This civilisation is finished'*[107]. Far from receiving a backlash, the article was praised. He therefore decided to put his name to it in the form of a speech at Cambridge University in October 2018[108] (he has since expanded on his theory in a book on the subject)[109]. In his speech, after a summary of the critical situation the world is in, he described three possible futures:-

- First up, we might *'transform civilisation'*. By this he means that we somehow re-configure the dominant 'Western', neoliberal, capitalist, hyper-individualist and unequal 'civilisation' we live in, so that it stops chewing up every last natural resource on Earth before spewing it all out again in the form of greenhouse gas emissions and any number of other toxic pollutants.
- Second, after *'some kind of collapse'*, a successor civilisation or successor civilisations might emerge; or
- Third, we dive rapidly into *'total collapse'* as global warming spirals out of control.

He has since labelled these three scenarios 'butterfly', 'phoenix' and *'dodo'*. The radical elements of the environmental movement have been dealing in the first possibility for decades, and largely still are. It tries to transform, restructure and re-engineer civilisation. Many have been pinning their hopes on this working and Read empathises with them when he says:

> *I hope that [a transformation of civilisation] happens and -*
> *probably like many of you - I'm actively working to make it happen.*

Read is sincere when he says this: a butterfly sits at the centre of the logo of the emerging TrAd movement he is leading. According to Read, TrAd is *'Win, win, win: we mitigate the effects of dangerous climate change, we work with Nature not against her, and we transform society in the direction it needs to transform anyway'*[110]. The analogy is clear, civilisation in its current form is an ugly caterpillar, but it could – if put on a nutritious diet – transform into a beautiful butterfly. The transformation he is talking about involves so much more than 'greening' the energy and transport systems. It involves fundamental changes in the way human beings relate to each other and the natural world. He is still clear that this is a hope and a goal, but not necessarily an expectation.

What Read went on to say in his speech challenges the wisdom of boxing ourselves into an assumption that in some way or other the climate and ecological crisis will be managed or 'solved', or that civilisation will be transformed. While hope is necessary, there is a growing need to consider what happens if we don't succeed. Read said this:

> *What I want to put to you this evening, among other things, is the thought that it would be a bold person who was prepared to commit to the thought that [transformation of civilisation] is going to happen. That we're going to make it happen, and that we're going to make it happen quickly enough. It would be a very risky bet to bet everything upon that kind of completely unprecedented transformation, and on overcoming all the vast vested interests and ignorance's and stupidities and laziness's and so on and so forth which stand in the way of it. For such a bet would occlude attention and resources starting to be devoted to take seriously the question, 'What if we fail?' How then could we make things less bad for whoever follows us?*

'What if we fail?' is a short but enormous question. It asks us to contemplate a bigger one: 'is this civilisation going to fail?' For many, that question sounds ludicrous. As ludicrous as asking: 'will the sun rise tomorrow morning?' 'Western' civilisation is a very solid construct in many people's minds. It is something people believe in and want believe in - it is a paradigm. What Read, Bendell, and others are arguing is that it is perhaps not as solid a framework as it seems to be. It is only by questioning its solidity that we can begin to fully contemplate the *'what if we fail?'* question.

So, is this 'civilisation' going to end? Maybe. The next section explores the signs that it might and how, politically and emotionally, this is being handled. It is then possible to ponder Read's follow-up question: how (if we fail) can we make things less bad for whoever follows us? In other words, what form might successor civilisations take? That is the subject of chapter 10, but first let's play whack-a-mole.

## Whack-a-mole

'Western', neoliberal, civilisation has been – for its winners – a relatively stable world. The number of winners, however, is comparatively low. Only perhaps a billion people, quite possibly fewer, are comfortably off. Life for the remaining six billion, not to mention the vast majority of plant and animal species, is less so and there are varying degrees of discomfort. Although the number is slowly falling, 10% of the global population live in extreme poverty - on less than US $1.90 per day. 3.4 billion people, around 46% of the world's population, are living on US $5.50 or less per day; and a staggering 85% of the world's population live on less than US $30 a day[111]. And yet the paradigm of 'Western' civilisation somehow prevails. It prevails because the promises it makes (but rarely keeps) are very seductive. For both those living *within* it, and those living *with* it, life in the comfortable version of 'Western' civilisation is *the* aspiration; it is *the dream*, (the *American* dream).

This dream does not grip us all, but it compels very many of us to consciously, or sub-consciously, ignore the signs that the 'Western' civilisation of our imaginations might be breaking down. We struggle, for example, to accept that most progress within it is slowing down[112], we turn a blind eye when we see the ugly consequences of its economic growth obsession, and we struggle to explain why wealth fails to do what it is supposed to do: 'trickle down'.

Signs of collapse are not easy to accept. The novelist Amitav Ghosh complains that climate change *'is like death, no one wants to talk about it'*[113]. Climate scientist Joelle Gergis notes how *'we are afraid to have the tough conversations that connect us with the darker shades of human emotion'* and cites TS Eliot's observation that *'humankind cannot bear very much reality.'*[114] Opportunities to question the foundations, assumptions and frailties of 'Western' civilisation are routinely missed. Even, it seems, in the midst of a catastrophic Coronavirus pandemic. This instinct to ignore and dismiss signs of collapse is a coping strategy: the temptation to swot them away, brush them under the carpet and pretend they don't exist is strong, it is a psychological adaptation. But the signs keep popping back up, making them harder to ignore, or keep hidden. Defenders of the status quo – who prefer the signs to be hidden – are forced to engage in a game of global whack-a-mole. They must either 'fix' problems as they emerge or divert our attention away from them. 'Western' civilisation will survive and spread for so long as it is possible to whack all the moles.

Today, however, the moles are popping up more and more often. They have sharper teeth, hairier heads, and redder eyes - they are increasingly angry. It is getting very hard to whack them all down. Their appearance in ever greater numbers might be the signal that 'Western' civilisation is nearing its end. COVID-19 gave it a good buffeting, 3°C of climate change could finish it off.

To recap, Rupert Read says: *'possibility number two is a successor civilisation after some kind of collapse [of 'Western' civilisation]'*. His conclusion is that this *'is what we have to start to think is likely to happen'*. Read wants us to *'think seriously about the successor civilisation'*. But before we do this, we need to explore the nature of 'Western' civilisation and what is meant by *'some kind of collapse'*, so that we can go onto explore how both relate to the birth of *'a successor civilisation'*.

---

The COVID-19 pandemic feels all-encompassing, the threat of climate change is monumental, and yet – right now – a short, sharp, one-off global catastrophic collapse event does not seem likely. What seems more likely, but by no means inevitable, is a series of calamitous events at district or nation state level that coalesce into a continental scale disaster. These individual calamities are the angry moles that keep popping up to warn us that 'Western' civilisation is in peril. But what exactly are these calamitous events? Filmmaker Adam Curtis opened Hypernormalisation, his epic 2016 documentary, like this:

> *We live in a strange time. Extraordinary events keep happening that undermine the stability of our world. Suicide bombs, waves of refugees, Donald Trump, Vladimir Putin, even Brexit*[115].

We can add COVID-19 to this list, but also previous pandemics like Ebola or SARS. We can also include the devastation of climate change events like Cyclone Idai, or the wildfires that have raged in California, Portugal and Australia. Calamitous events have always happened, but they are happening more often now, increasing in frequency and magnitude. The trouble is, as Curtis goes on to say: *'those in control seem unable to deal with them'*. Every time yet another calamitous event occurs, we ask ourselves why it happened. But with calamitous events of different shapes and sizes happening more and more often, it is getting harder to avoid asking the deeper question of why they *keep* happening? This more profound question can be acutely uncomfortable.

It forces us to ask searching questions about value systems and political structures that are usually left unquestioned - it's the difference between concluding that (a) the Chernobyl disaster was caused only by human error and faulty design within the nuclear reactor; or concluding that (b) it was human error, faulty design *and* the failings of the Soviet system. Conclusion 'b' was an immeasurably more profound one to reach for supporters of the Soviet Union. Asking profound questions that expose brutal truths are also uncomfortable because (Curtis again) *'no one has any vision of a different or better kind of future'* - this is one of the key reasons those difficult questions are rarely asked.

Of course, there are many people who see events unfolding and think 'oh dear', move on, and don't let it trouble them too much[116]. There are others too, as Elizabeth Sawin explains[117], who believe so strongly that they *'live in the best possible way'* that it is psychologically impossible for them to see the increased frequency of calamitous events for what they are - a sign that maybe we don't. Curtis builds on these themes in the introduction to Hypernormalisation, which is split into three quotes here:

> *This film will tell the story of how we got to this strange place. It is about how over the last 40 years, politicians, financiers and technological utopians, rather than face up to the complexities of the world, retreated. Instead, they constructed a simpler version of the world in order to hang onto power; and as this fake world grew, all of us went along with it because the simplicity was reassuring.*

His theory, inspired by the work of Alexei Yurchak[118], is that because reality is too complex or uncomfortable to contemplate, Westerners have decided to pretend, for as long as they can get away with it, that nothing is wrong. Yurchak argued that this is what happened in the Soviet Union as it drew closer to collapse. Curtis argues that it is happening now in 'Western' civilisation.

> *Even those who thought they were attacking the system, the radicals, the artists, the musicians and our whole counter-culture actually became part of the trickery, because they too had retreated into the make-believe world, which is why their opposition has no effect and*

> *nothing ever changes.*

Their opposition, like the efforts of much of the mainstream of the environmental movement, dealt only with surface level change. Deeper transformation was aspired to, but always harder to imagine, harder to achieve, or crushed before it gained any momentum.

> *But this retreat into a dream world allowed dark and destructive forces to fester and grow outside. Forces that are now returning to pierce the fragile surface of our fake world.*

The 'forces' are the fuels that power the calamitous events that keep happening. They give birth to the angry moles that *'pierce the fragile surface'*.

Whilst 'Westerners' have been riding along in a *'make believe world'*, global heating, racism, extremism, authoritarianism, infectious diseases, inequality and more have been growing in a murk that many people prefer to leave unexamined. They gladly leave it in the periphery of their daily experience - signs of collapse are not easy to accept.

Many of us, in the West, might therefore need to accept that we are more than just innocent bystanders who merely witness calamities. We might, in many ways, be complicit. In the case of greenhouse gases especially (but most of the other angry moles too), what we have been doing to build and enjoy our *'fake worlds'* has directly fed the destructive forces that are now manifesting themselves as painful – human induced – calamitous events.

Until now, Western governments have managed to contain most of the angry moles. The combined strategies of whacking them (military intervention, emergency aid); obscuring them (propaganda and the creation of 'fake worlds'); and denying responsibility for them (climate scepticism and consistent othering) have worked and *kept the show on the road.*

If a *getting worse, then getting even worse* scenario turns out to be the one that is ahead, then the world stands on the brink of catastrophic global heating – something that could be very difficult to adapt to or reverse. If that *is* the case, it is game-changing. At that point, climate change itself has changed – tipping points will be passed, leading to cascades of change in the Earth's system – it will take on its own momentum. In this intensifying form, climate change will mix with other forces (that are also intensifying) and multiply. Together these forces, this growing population of angry moles, has the potential to trigger 'local' calamitous events that are even more frequent and even more intense all over the world. The great global whack-a-mole game will have been lost, 'Western' civilisation will have had its 'Chernobyl' moment.

To some, this seems to be inevitable. What other outcome can there be if the fuel keeps flowing to the destructive forces and they keep mixing together to form a toxic, explosive soup? In the worst-case scenario, these events could become so big and so frequent that they might even coalesce into a superstorm of absolute chaos.

# 9. Different or better kinds of future

What is there left to hope for? Are there any *visions of different or better kinds of future* that might emerge from a collapsed or disassembled 'Western' civilisation? Will a wonderful 'Phoenix' rise from the debris? Or could *visions of different or better kinds of future* inspire a transformation away from 'Western' civilisation *before* it turns to dust? Is there a beautiful 'Butterfly' waiting to grow out of the consumer capitalist caterpillar?

There has never been a careful disassembly of anything as big as 'Western' civilisation, so it is hard to find many butterflies to be inspired by. However, there have been calamities and collapses, and phoenixes have risen from them. Some phoenixes are less wonderful than others (as Naomi Klein made clear in her book 'The Shock Doctrine'[119]), but clues as to what futures might be possible can be found by studying the forms of society and democracy that are rising from the aftermath of contemporary calamities and collapse events.

Of course, one key consideration for those envisioning possible futures is to question whether the search is simply for *the* alternative, or *the* successor to 'Western' civilisation. It may be that this singular way of thinking about civilisation is part of the problem - it is, after all, a seed from which crusades to *civilise* and *colonise* others can grow. It may be better to think more plurally and be open to the development and envisioning of alternative and successor civilisations - the many ways we humans can (and could) interact with each other, other species, the built and natural environment, and the climate. Thinking pluralistically and resisting hegemonizing forces creates the potential for the emergence or (re-emergence) of *many* civilisations of varying scales and diversity.

---

The war in Syria, by any measure, has been horrific, life has collapsed for millions of people in one fell swoop, it has been an utter catastrophe. This was more than a calamity, it was a collapse, and it was driven and exacerbated by at least two, if not three, angry moles. It was rooted in an uprising against the authoritarian rule of President Bashar al-Assad, which had been building for many years - pro-democracy protesters had been piercing through the fragile surface of Assad's regime for some time.

It is thought that another angry mole, climate change, played a role too. A period of intense drought, which caused food shortages and internal migration created the conditions for wider protests, and then fighting, to break out - desperate times call for desperate measures[120]. Then, once the war was underway, ISIS - a third angry mole - seized its opportunity. They spread across Syria to expand their violent rejection of (amongst other things) the forces of global hegemonic 'Western' civilisation that were arriving in Syria in the form of US and other Western militaries to fight Assad in support of Syria's pro-democracy movement.

As the tail end (hopefully) of the war in Syria approaches, it will be important to examine what happens after the collapse it has experienced. The players in that war (a combination of Assad, Russia, the US, the UK and the various Syrian rebel armies who fought ISIS; but also those Syrian's who favour none of the above)

## DIFFERENT OR BETTER KINDS OF FUTURE

are now all engaged in a competition to shape (or re-shape) post-war Syria. The Americans no doubt want the 'global hegemonic civilisation' to take root, Putin will want something slightly different, Assad something different again, Erdogan will have his view, and ISIS may continue to have their say.

What of the people themselves? In Rojava, the Kurdish region of northern Syria (or, depending on your perspective, the Syrian region of Kurdistan), a movement has been building: *'Make Rojava Green Again.'*[121] It is an attempt to build an ecological society, based on grassroots democracy, women's rights and linked to the teachings of Abdullah Öcalan and Murray Bookchin. It is something like the 'Phoenix' that Rupert Read describes.

The movement started while the collapse (the war) was happening. The retreat of any form of functioning national or local government in Rojava had created a vacuum. With US and Coalition air support, local militias fought to ensure ISIS didn't fill that vacuum. As the local militia held the region, the political void was filled with grassroots anarchist organising. This is anarchism in a political sense, a deliberate form of governance. In Rojava, anarchism has given birth to what is now known as the Autonomous Administration of North and East Syria. The feature length documentary *'Accidental Anarchist'* by Carne Ross[122] provides a useful introduction to its origins and history.

The 'Make Rojava Green Again' movement has strong ecological, multicultural, democratic, and feminist principles. It is based on a political system of democratic confederalism, where power is devolved to as local a level as possible, with decisions deliberated

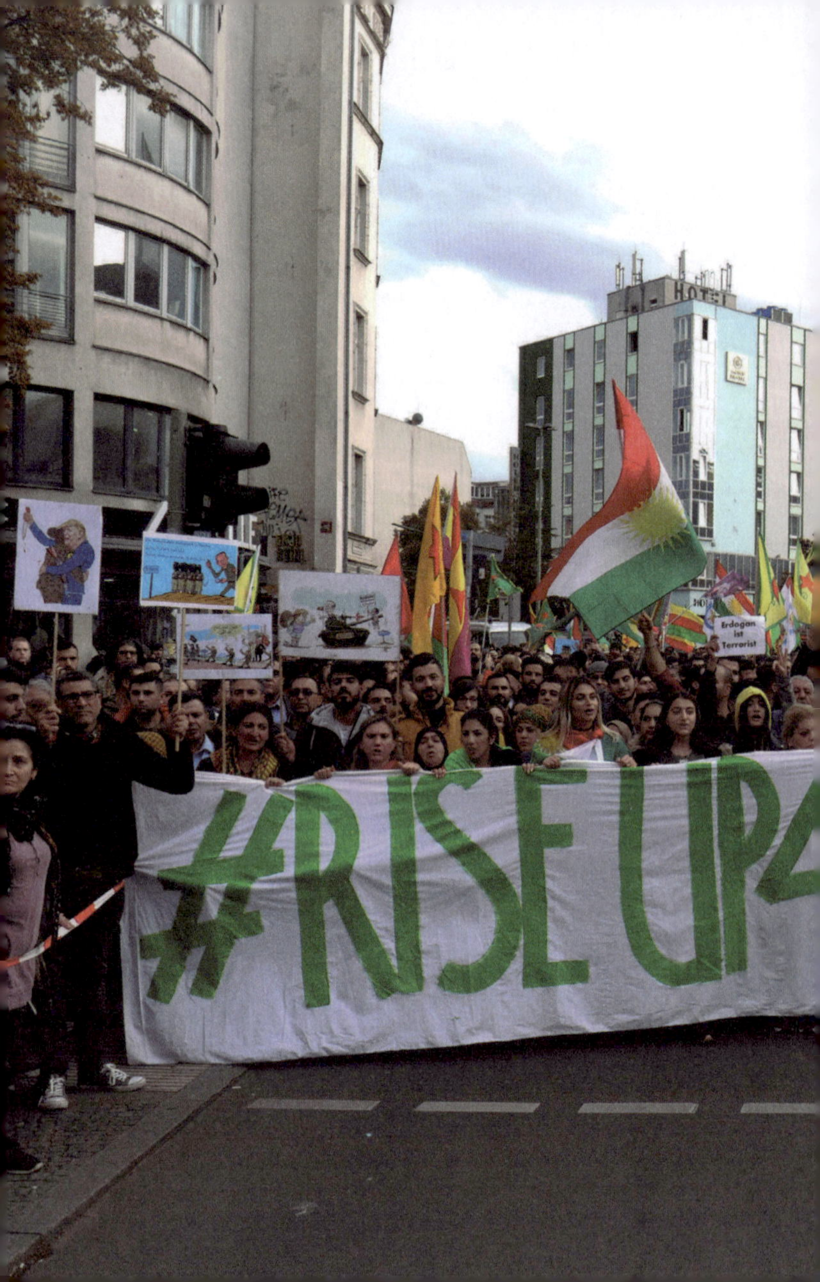
HOTEL
Erdogan
ist
Terrorist
#RISE UP

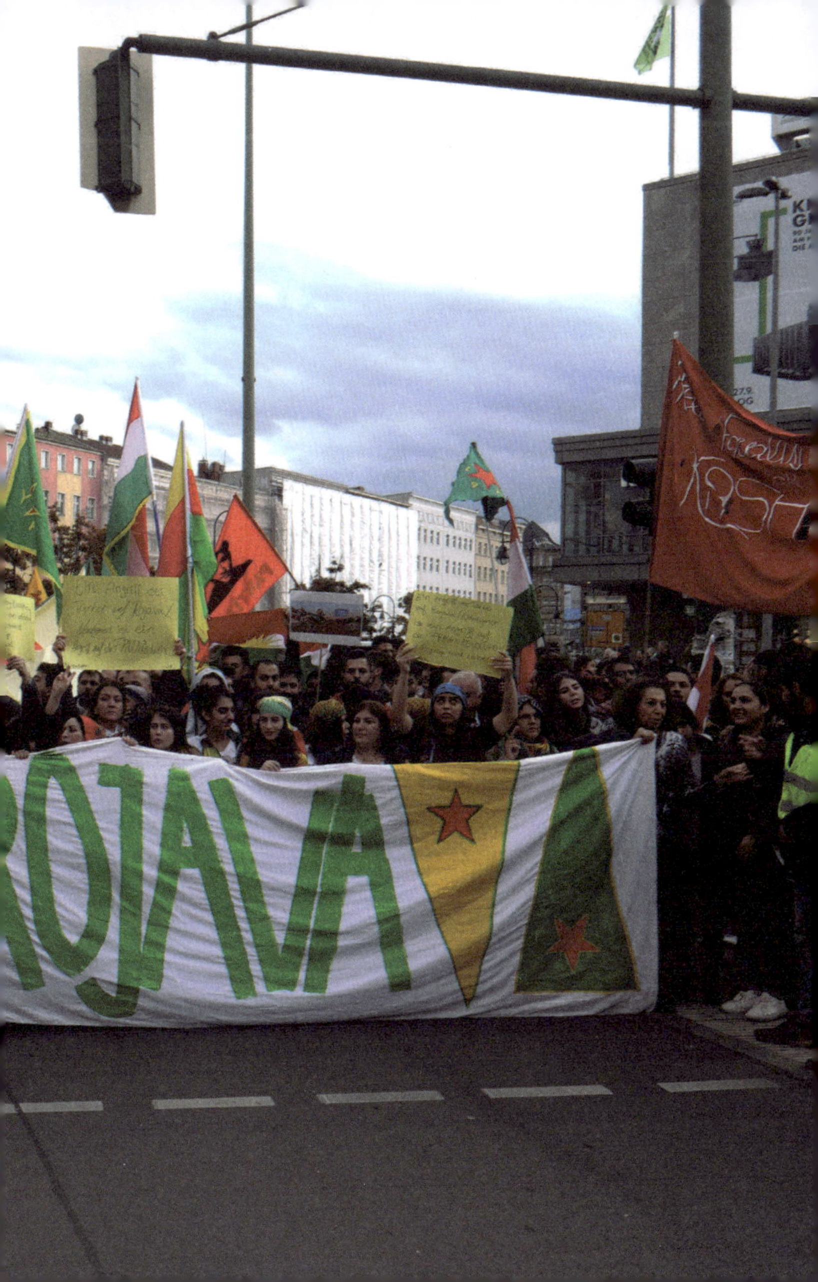
ROJAVA

through citizens' assemblies, and resources allocated through participatory budgeting. Reforestation, renewable energy projects, lifelong learning, agro-biodiversity, and mutual aid are all present and flourishing. It rejects what it deems to be unhealthy about the global hegemonic civilisation (but adopts some of the better bits), but it also rejects the ISIS worldview, and seeks to keep Assad, Erdogan, Putin, et.al., at arm's length. It is not perfect, but it is genuinely different, and it is not uncomfortable to imagine being part of it (or something similar). It looks like one of a number of potential 'successor civilisations' or successor societies that might be emerging.

Now, there is no guarantee that the 'Make Rojava Green Again' movement will succeed, and it is easy to overromanticise it. Rojavans will need strong diplomatic support, financial support and no doubt, some luck; the odds are stacked against them[123]. But it provides a vision and helps us see that 'Western' civilisation does not have to be the end goal of 'development'. 'Make Rojava Green Again', and other 'Phoenix' like it, are so important because they help us to imagine *different kinds of future.* Rojavan's are willing to challenge the value structures that underpin 'Western' civilisation. They take the good and reject the bad as they build their own model based on value structures that make more sense to them. In this way, they build strength against destructive forces that they are not afraid to face and build Deep Adaptation style resilience to future calamities that will undoubtedly hit them.

However, for every hopeful and inspiring example of a potential successor civilisation, there are hundreds of examples where the response to a calamity has been to try to rebuild everything to match the way it was before, or to impose something new that is worse than what came before. When a society experiences a shock to the system, it is common for individuals to want normality to return, and for it to come back as soon as possible; it is an almost instinctual response. In 'Western' civilisation, the wisdom of doing this seems to ebb away with every passing decade. If we keep rebuilding the same social, economic, environmental and political structures, we risk seeing them implode again and again. In places where 'Western' civilisation has not yet embedded, the task, after any future calamity suffered, is maybe to resist the disaster capitalists who, in countries like Syria, will be goggle eyed at the prospect of imposing 'Western' civilisation in places like Rojava.

When Kate Marvel wrote about needing *'courage, not hope, in the face of climate change'* this is maybe what she was talking about[124]. The Autonomous Administration of North and East Syria is the result of great courage. Its founders have fought physical battles of incredible hardship and are now fighting an ideological battle to achieve official recognition within Syria and to 'Make Rojava Green Again'. It is an existential quest and has strong echoes with what Marvel famously said: *'courage is the resolve to do well without the assurance of a happy ending'*.

'Make Rojava Green Again' tackles climate change through ecological intelligence; it addresses religious extremism by seeking to reintegrate rather than lock up those who cooperated with ISIS; it questions the global hegemonic civilisation; it exposes the very idea of imperial 'civilising' agendas. It does these things simultaneously. Rojava isn't being 'civilised' by someone else, it is being built by its people, for its people. It has taken advice and support, but it hasn't taken orders. If it blossoms, it could be a working example of a 'successor civilisation' that has succeeded.

## Are collapse events necessary precursors to successor civilisations?

Learning about Rojava left me wondering if a calamity or collapse *has* to happen before a successor civilisation can emerge? Surely not, ideally not (I'm not a collapse fetishist), but it might be so, so the question needs to be confronted. Clearly, we really, *really* want to avoid calamities and collapse events, but they will happen, so let's look at that scenario first. At least two things happen when such an event occurs:

> Firstly, severe calamities strip the fabric away. They leave a vacuum, a blank canvas; they create a space for something new. This is most likely to be the case for events like wars, cyclones and forest fires. But mostly wars, because they create political vacuums - as is the case in Rojava.
>
> Second, those caught up in calamity events encounter reality in a very tangible way - not via a screen. Those who survive have beheld an existential threat. Do such visceral experiences make the difference? Do they lead people to commit fully to a slogan like 'never again?' Have they been through so much with enough people

that they form the critical mass and solidarity needed to start something new? Do calamities act as a trigger that leads to reinvented futures?

Collapse events *can* create conditions from which successor civilisations might grow - like a phoenix from the flames. But it would be Neo-Malthusian in the extreme to sit and wait around for them. As much one might romanticise Rojava - or idealise about other utopian successor civilisations that have arisen out of the chaos of war and disaster - to stand-by and let preventable disasters happen in the hope that utopia might follow is obviously abhorrent, and unnecessary. Calamities and collapses are categorically not required precursors to transformational change, humans have restructured the way they organise themselves in thousands of ways over the centuries. There has always been an element of struggle, but shifts from the old to the new have happened and go on happening in calm and considered ways.

The second scenario - a society transforms peacefully to entirely new social and economic structures - is possible. Ugly caterpillars *can* turn into beautiful butterflies. They don't need the crumbled foundation of a calamitous event. Brownfield sites beat bombsites every single time, Read's vision of 'transformative adaptation' must surely be possible.

In Rojava, the leaders of the Autonomous Administration of North and East Syria are making the case for their approach. They need international diplomatic support to help them convince Damascus to give them official recognition in a federalised Syria.

In seeking this support, they are trying to prove that their approach is bringing peace and argue that Rojava could become a beacon of stability in the Middle East. Time will tell of course, but if Rojava is successful, elements that are helping it succeed - like their citizen's assemblies - might be replicated elsewhere. If this happens, it could indicate two things. First, neighbouring states in the Middle East might be adopting the Rojava model out of a desire to dampen down a volatile situation of their own. They may see Rojava achieving a lasting peace and decide to follow their lead to bring stability to their own situation before any current fragile state turns into something more dangerous and violent. Secondly, adoption could be the result of admiration. It might be that they observe the Rojava model and like what they see. They would then study it, trial it, adapt it to fit their own locality and customs, and then strive to make it work.

---

In any of the potential climate change scenarios that lie ahead, it is likely that successor civilisations will emerge with and without the trigger of a calamitous event. The hope of course is that the pioneers provide great examples that others can learn from and implement in peaceful ways - before calamity hits. But be under no illusion, there are reasons why there are so few examples of the sort of politics that is emerging in Rojava. Carne Ross explains a key one:

> *Those who have power have a strong interest in retaining it and have done a lot to suppress alternative modes of the economy or politics. Power is a zero sum. We can't all be more powerful. If*

> *people at the bottom are to be more powerful it means people at the top have to lose power and people don't like giving up power.*[125]

There is however, the potential for leaders to emerge who seek to gain power so that they can give it up. These sorts of leaders would support deliberative democracy of the *'people's inquiry on climate change'* sort discussed in chapter 8. They would, in fact, be more ambitious than that and enable places like Warwick to run full citizen's assemblies on climate and numerous other local, regional, and national issues. They would be empowering citizens to not just make statements and recommendations that guide the decisions of elected representatives, they would be trusting citizen's assemblies to actually make the decisions, as per Rojava.

The Rojavan model is still embryonic, and citizen's assemblies still need to evolve. But the Make Rojava Green Again movement might one day be looked back on as an early example of a blended version of politics that is teething its way through. It has the potential to show that collectivism and individualism can – counter to expectations – co-exist. The democratic confederalism being practiced in Rojava allows people to express their individualism in a way that makes acting collectively part of that expression. It works because citizens are enabled to have a role in political decision making - they contribute their thoughts, experiences, and feelings to the debate in a meaningful way. Having had that opportunity, citizens are more accepting of the decisions made, and more invested in making them work. It is a form of governance that aligns with *survival of the friendliest* framings of human nature. It sees people as creative, compassionate, and cooperative citizens, and wants to empower them. It does not see them as selfish, instinctual, and destructive consumers that need to be controlled. It is not scared of individualism, it accommodates and uses it - it values the ideas and perspectives that individuals offer. The policies that emerge through this process are therefore richly informed, and more effective.

Outside of Rojava, it may be that both the Left and the Right have struggled to inspire transformation because too many of their thought leaders are trapped in the *survival of the fittest* paradigm. They mistrust human nature and that mistrust spreads through society, hindering the prospects of mass civic action and the cooperation that has made the human species so successful[126,127]. What has been developed in Rojava differs so fundamentally, it is a vital antidote to the stale and tired visions of the future served up by today's political leaders.

What *Make Rojava Green Again* might teach the world is that to formulate visions of *different and better kinds of future*, we may first need to re-examine our deepest assumptions about the values of our fellow humans. And this may lead us to temper any concerns we have about the rise of individualism.

Many environmentalists see individualism as a barrier to the sort of collective action that the climate and ecological crisis demands of us. Collectivism, however, is something individualists fear - it is a ghost from the past and they defend individualism as a safeguard against it. These are the tensions that exist – they have been there for centuries – but right now, belief in the individual as the unit of greatest importance is stronger than it has ever been.

Individualism is far from perfect – it can mutate into toxic selfish forms – but it does still have merits, for example: freedom of expression, creativity, autonomy, and various other forms of liberalism. These are worth defending, and people do - most arguments and injustices can be traced back to one individual's pursuit of freedom clashing with another's. But this is not inevitable.

It pays to remember that today's version of individualism is in its very early stages, it has a long way to develop. If it were a human being, it would be a toddler. Right now, it is as likely to display acts of deep and authentic compassion, as it is to throw a massive temper tantrum. But one thing individualism is not doing, is going away. The question is how will it mature? Will individualism grow up to be an unruly teenager and dysfunctional adult? Or will it blend in some way with collectivism, so that individuals can simultaneously identify as a unit of one, and a unit many? Deliberative democracy is growing in popularity around the world[128], this is a very promising sign. Individualism and collectivism could find a way to merge into one another.

# 10. A note from the author

Here's where I'm at: I still believe it is possible to prevent catastrophic climate change, but only if we transform the way our societies and economies work. I believe that we can create these transformations and that we will when inspiring visions of *different and better kinds of future* come into focus.

These visions are coming into focus for me, they have required me to think and work in the *'personal'* sphere of what Karen O'Brien has categorised as the *'three spheres of transformation'*[129] and not get bogged down in the inner spheres of *'practice'* and *'politics'*. The *'personal sphere'* is where paradigms, worldviews, beliefs and values are made and re-made by individuals and communities. Transformation in that sphere is what creates the possibility for the transformation of political structures and the practical projects that flow from them.

In recent years my personal sphere has been most heavily influenced by the work of Hilary Cottam and Anne-Marie Slaughter on 'Sapiens Integra'[130] and the paradigm shifting analyses of human nature that Rutger Bregman,[131] Brian Hare and Vanessa Woods[132] have recently popularised so brilliantly. I am also indebted to my colleagues in the NGO sector, specifically Tom Crompton at the Common Cause Foundation[133], Rob Bowden and Rosie Wilson at Lifeworlds Learning[134], Jon Alexander at New Citizenship Project[135], and my brilliant colleagues at Global Action Plan[136]. Books like 'Less is More' by Jason Hickel[137], 'Doughnut Economics' by Kate Raworth[138], 'Hyper- Capitalism' by Larry Gonick and Tim Kasser[139], 'The Next Revolution' by Murray

Bookchin[140], 'Leaderless Revolution' by Carne Ross[141], and 'Make Rojava Green Again', are helping me to imagine entirely different political structures and how life, in a practical sense, could look if transformation can be made real.

Further hope and inspiration can be found in books like 'Finntopia' by Danny Dorling and Annika Koljonen[142], in the political choices made by countries like Costa Rica and New Zealand, in the writings and imaginations of Debbie Bookchin, Rebecca Willis, David Graeber, Martin Kirk, Satish Kumar, Giorgos Kallis, Mariana Mazzucato, Nathan Thanki, Amy Westervelt, Skeena Rathor, Julia Steinberger, Andrew Simms, and in the visions and values of the new wave of progressive political figures who are emerging all over the world.

There are examples in this book that hint at transformative change, researching them filled me with hope and ideas. I am excited by what is happening in Warwick, Glasgow, and Morocco, and by the conversations that bounce around the 'TrAd' network. The work being done by the Climate and Migration Coalition to re-frame and enable migration is truly inspiring. And I am also, of course, incredibly fortunate to be involved in the community-led agroforestry movement in rural Nepal - it is a *demonstration* of a different and better kind of future; my colleagues at EcoHimal and HICODEF are heroes.

So yes, I am worried about the future, but I have not given up on it. I believe that we can give birth to better futures, but I understand that there is a gestation period and that it won't be pain free. There is a lot of work to do.

The eco-philosopher Joanna Macy has said this: *'It looks bleak. Big deal, it looks bleak.'*[143] I used it as the title for chapter 8. She's saying it to people like me. I am a 41-year-old white male with a decent salary, a loving wife, newborn son, and a comfortable home. When I study the climate and ecological emergency, it certainly *looks* bleak, and I do have deep concerns about the state of the planet. But for me, in my position of privilege, it isn't yet *actually* bleak.

Macy is urging me not to wallow in the bleakness, not to join the *'climate sad bois'* brigade that Kate Aronoff rightly scolds[144]. Whenever I start floundering in the enormity of it all, I need to accept that yes, it is bleak, yes, it is sad, but then I need to get over myself - this isn't about me, I'm not on the frontline, and I have a lot of agency. What's required according to Macy is this: live in the moment, get stuck in, connect with those around you (humans and non-humans). Go on living and loving, creating and thriving - but in different ways. She wants us to transform our ways, but not by a bit, by a lot. I agree with her.

This translates into hoping and working for a *'butterfly'* future, while not ruling out or fearing a *'phoenix'* future. As for extinction of the human species due to climate and ecological breakdown, that is not going to happen in my lifetime, not in my son's either, nor in my grandchildren's or great grandchildren's. I honestly don't think it is possible. Human beings are too cooperative, too brave and have too much ingenuity to let that happen. Our numbers might peak and then go into decline and there will be enormous loss and suffering, but climate and ecological breakdown won't kill human life off. Our species won't die out, it will survive and likely thrive in real but imperfect successor civilisations. We will make *great* and *just* adaptations to whatever lies ahead, that is my hope.

# Part 4 – Stories

> *Far from signifying climate ambition, the phrase "net zero" is being used by a majority of polluting governments and corporations to orchestrate escape clauses so as to evade responsibility, shift burdens, disguise climate inaction, and in some cases even to scale up fossil fuel extraction, burning and emissions. The term is used to*

greenwash business-as-usual or even business-more-than-usual. At the core of these pledges are small and distant targets that require no action for decades, and promises of technologies that are unlikely ever to work at scale, and which are likely to cause huge harm if they come to pass. ”

*Global Campaign to Demand Climate Justice* [145]

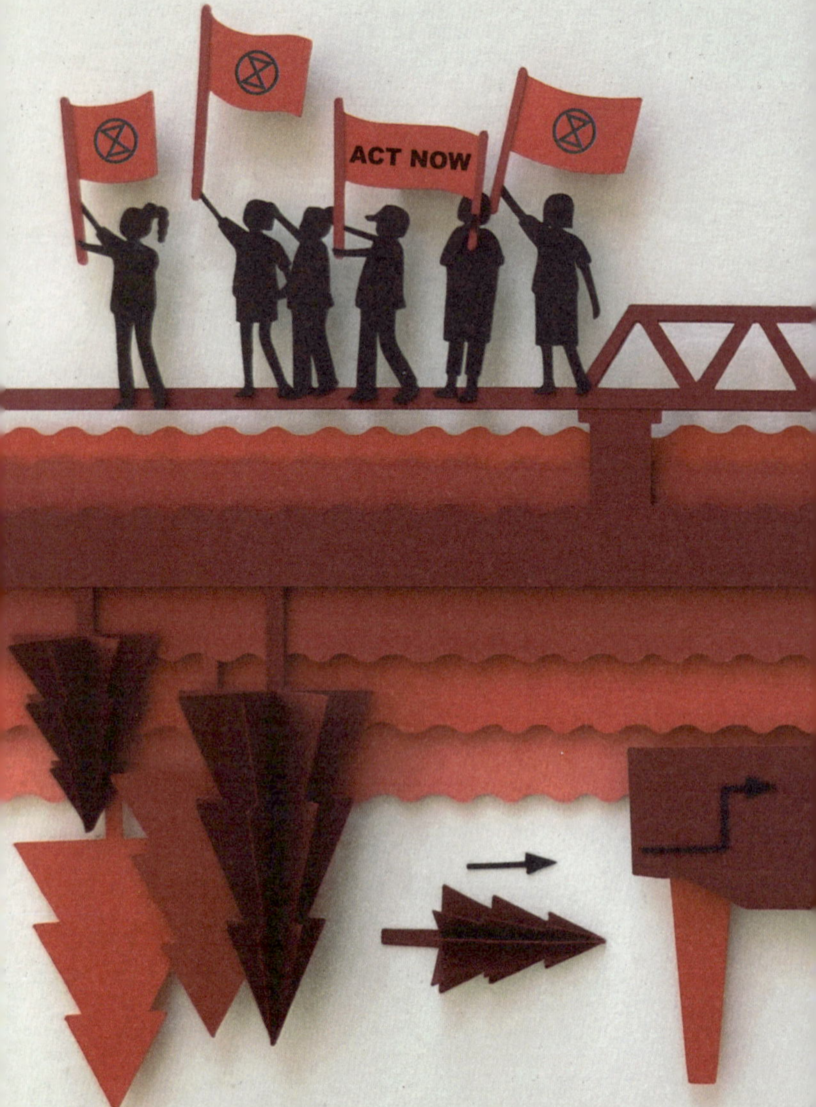
ACT NOW

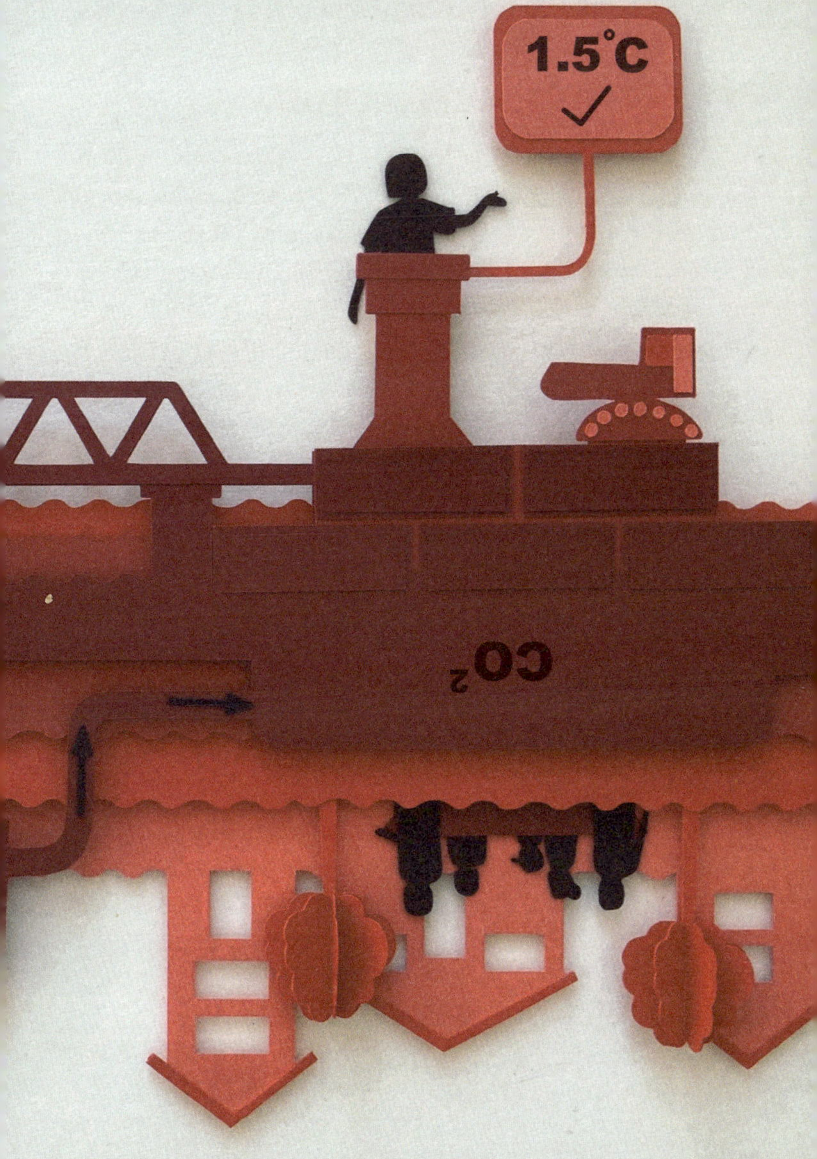
1.5°C
$CO_2$

# 11. The Reassuring Story

Chapter 9 introduced the idea that the UK Government likely knows that even if the UK successfully achieved its *'Net Zero by 2050'* goal, it will be one of only a few countries that does. The UK's efforts would therefore be, not futile (the right thing to do, is the right thing to do, and every tenth of a degree of warming that can be avoided is worth avoiding), but certainly of limited effect at a global level. The truth is that while some nations (e.g. Norway, Sweden, Finland, and New Zealand) have even more ambitious targets than the UK's, there are many more countries, including some super-sized ones, who are yet to even set a Net Zero target, let alone get themselves on a credible path to achieving it. But what of the UK's path to Net Zero - is it credible?

Putting aside concerns about *'Net Zero by 2050'* being the right target (it is very likely too little, too late[146] and a smokescreen), there are serious doubts about the credibility of the Government's roadmap for achieving it. It is clear that they are making assumptions about the future, and about future technologies that are at best optimistic and at worst dangerously fanciful. The flaws in their plans have been pointed out multiple times by Caroline Lucas MP[147], Kevin Anderson and colleagues[148], various NGOs and campaigning bodies[149], and many of the world's leading climate scientists[150], all of whom have serious doubts on scientific (not to mention political) grounds about the Government's ability to back up their rhetoric with the action that is required.

Government figures are surely not deaf to these concerns. In private, Ministers and their allies in business and civil society may well be admitting to themselves and each other that the UK is not on track to achieve Net Zero by 2050. This is not something they can or will admit in public, because if they did it would destroy a powerful and reassuring story about the future that they are determined to keep society wrapped up in.

The reassuring story is one in which Government and big business have climate change under control. It is a nice one to hear because it paints *business-as-usual* as viable, sensible, and compatible with the prevention of 1.5°C of warming. It is a comforting story of small lifestyle tweaks, shiny technologies, profitable 'green' investments and incremental change. It no doubt reassures the Ministers who parrot it out as much as it reassures anyone who merrily goes along with it.

Few people question this reassuring story, not even opposition shadow Ministers. There are large sections of the environmental movement that go along with it too, in public at least. They're often compelled to by peer pressure, expectations of supporters, colleagues, and funders. They have to be seen to believe in it even if they don't - questioning it has consequences.

To question the *reassuring story* is to open Pandora's box. It forces one to ask if the Net Zero by 2050 package of measures is a *'present which seems valuable, but which in reality is a curse'*[151]. It also makes one confront the possibility that climate change is *not* under control and whether *business-as-usual* needs an eyewatering overhaul. That is a deeply uncomfortable set of thoughts to contend with. It is especially unsettling in a world that already feels turbulent and unhinged - little wonder keeping a lid on the box marked *'reassuring story about climate change'* feels preferable.

However, knowing what is in the box is important, so it needs to be opened - even if it is scary.

Judging by what various Government arms are busy doing in the background, it seems that they know what might and, more pertinently, what might not be available to us. As packages of measures go, it is quite underwhelming. There isn't much in the box to get excited by. The Government almost certainly knows that 1.5°C is an extreme longshot, that 2°C is on a knife-edge, and that 3°C and 4°C are both distinct possibilities. It is with all this in mind that the UK's official 'Climate Change Committee' conducts its five-year climate change risk assessment. The most recent one was published in summer 2021[152] and assessed the implications for the UK of average global temperature rises of 2°C and 4°C. These risk assessments are the basis for the Government's official National Adaptation Programme (NAP)[153], which every signatory of the 2015 Paris Agreement is required to produce and periodically update. It is of course a good thing that these NAPs exist, but few people are aware that they do. This is because the story the NAPs tell is very much *not* part of the reassuring story, NAPs tell of a world in which *business-as-usual* is clinging on for dear life.

Few people want to hear that story, even fewer want to tell it, especially those in, or close to power. It is because of this that the launch of the UK's next NAP (scheduled for 2023) will come and go with little fanfare. It won't be on the front pages, and it won't become common knowledge that the Government is making serious plans for a world that is 2°C, or even 4°C, warmer. The reassuring story will continue to drown it out.

Given the robustness of the reassuring story, it is no wonder that great institutions like the RNLI were not prioritising climate change adaptation as much as they probably should have been when we first spoke to them. The RNLI was probably, like so many others, feeling that reassurance. They were assuming that climate change, despite being a threat, was under control. Fortunately, following a change of CEO (and I hope, in part, because of the nudge the four-page report The Glacier Trust sent them on the risks posed by climate change), the RNLI is now consulting with scientists and its key stakeholders to develop a full climate change adaptation plan. A first draft is due by the end of 2021, with the final plan to follow soon after.

Sadly, however, the continued predominance of the reassuring story (that rarely mentions the need for adaptation, let alone transformation) means that the RNLI is an outlier in making serious preparations to adapt. Other social actors – from individual families, communities, ski resorts and wine growers, right up to multinational companies – are more relaxed, or possibly naïve. They either have no adaptation plan at all, or a very basic one that assumes that climate change is under control. How many have a plan flexible enough to withstand a 2.5°C, 3°C, or 4°C rise? It is unlikely to be many.

---

I do not wish to overly criticise the motivation of the Government and its allies in sticking to the reassuring story. They have good reasons to. Having left it so late to act, the action that is now necessary to prevent a 1.5°C or even 2°C rise is very much in the 'drastic' category. It will involve far more than tweaks to business-as-usual, and that reality, as pointed out above, could unsettle a lot of people.

The enormity of what needs to be done should not be underestimated, and the challenges are not merely technical. They are social, economic, cultural, even philosophical. They are emotional too, there is a lot to wrestle with. Anything is possible, the world can be remade, but it will not be straightforward. To illustrate this, and to crack open Pandora's box a little more, it is worth taking a slight detour into the fantastical world of climate-saving big-tech.

You are probably used to hearing about the technological developments in renewable energy, hydrogen and electric vehicles, smart agriculture, eco-building, and electric food. Widespread scaling and mass adoption of them all (and others) is a central tenet of the reassuring story. Progress on many of these fronts will no doubt be achieved and will be a cause for celebration, but it won't necessarily be cost-free - a socially, racially and ecologically *just transition* from one set of technologies to another is not guaranteed. But there is another category of technological advancement that receives a lot less attention. This is despite it having almost mission critical status in the calculations that underpin the reassuring story; it is a central pillar. To put it crudely, plans to reach Net Zero by 2050, and limit global heating to 1.5°C, or *well below* 2°C, are reliant on an almost exponential growth in a suite of innovations, projects and appliances known as Negative Emissions Technologies (NETs).

In essence, 'NETs' is a catch-all term for the various ways in which greenhouse gas emissions can be removed from the atmosphere or prevented from escaping into the atmosphere at all. Trees are a negative emission 'technology'; so are 'scrubbing' machines that capture post-combustion carbon in a solvent to prevent it from being plumed out of a power station chimney. Guided by the UN and the IPCC, Governments all over the world have invested a lot of hope (if not yet a lot of money) in the potential of NETs. They are putting a lot of faith in technologies of various tested and untested forms. Confidence is even being put in NETs that have not yet been invented, or even imagined. Put bluntly, world leaders are putting a lot of trust in the ingenuity of scientists and engineers who are currently still in Primary school.

Whether or not faith should be placed in NETs is contested by climate scientists, modellers and social justice activists. Yet in the climate movement, let alone the wider population, there is very little awareness of the scale at which NETs would need to be deployed to have the desired impact. The plans for one key category of NETs – 'Bio-energy with Carbon Capture and Storage' (BECCS) – reveal the size of what is being dreamt up and the gambles being taken.

BECCS, in its most straightforward sense, involves growing billions of fast-growing carbon-sucking trees, harvesting those trees, transporting them to a power station, burning them to power an electricity generator, capturing the carbon dioxide ($CO^2$) from the smoke, and then burying it in liquid form in abandoned underground aquifers and oil wells. So far there is only one commercial scale BECCS facility in the world - the Archer Daniels Midland (ADM) plant in Illinois, USA. This facility doesn't actually burn trees or produce electricity, it burns corn to produce a fuel: bioethanol. ADM are able to capture 100% of the $CO^2$ which is then sequestrated in the Mt. Simon Sandstone formation - a nearby saline reservoir in the Illinois Basin. The ethanol, one of ADM's key commercial products, is then sold to large energy companies who blend it with unleaded gasoline to create a cleaner (but not clean) fuel for trucks, railcars and barges[154].

There are a few more BECCS facilities around the world, but none are working at a commercial scale. However, there are nearly twenty commercially operating facilities that do the Carbon Capture and Storage (CCS) bit. Unfortunately, these burn fossil fuels, rather than trees or plants. Another twenty or so BECCS facilities are planned, but it is scaling up slower than hoped, and much slower than required if it is going to be anything more than greenwashy sideshow.

Exactly how much scaling-up needs to happen can be illustrated by looking at what is being achieved today. Between them, the 18 CCS facilities that are currently online are capturing around 40 million tonnes of $CO^2$ per year, but only 10% of this is being stored in the planned geological way[155]. According to the models being used to plot a path to Net Zero, CCS – most especially BECCS – will become a 'solution' to the climate crisis, when it is capturing 15 billion tonnes of CO2 per year, forever[156]. A *lot* of trees and approximately 15,000 BECCS facilities are going to be needed to achieve this![157] To put that in context, the UN estimates that by 2030 there will be approximately 1,400 cities with a population of 500,000 people or more, and that 60% of the world population will be living in these cities[158]. 60% of

15,000 BECCS facilities, is 9,000 BECCS facilities. So, on average, every one of those 1,400 cities will need to find space for six BECCS facilities.

The local opposition to this is likely to be significant, these facilities are not small or pretty things. Imagine submitting a planning application to say, Birmingham City Council asking for permission to build six brand-new BECCS facilities? You'd need to be lucky (or very *well connected*) to get that through. Consider too, that six wouldn't even be a fair share for Birmingham (population 1.15 million), it would need to build 12 or 13 facilities. And yet, this is only one potential barrier to the mass scaling up of BECCS. Imagine how hard it is going to be to secure the land, water and fertilisers needed to grow all the trees and plants required. Land-grabs of any scale are met with opposition, a water and land-grab at this scale would be met with extremely fierce resistance - and quite rightly.

Almost unbelievably, BECCS is the NET that is right at the centre of the plans to prevent temperatures rising to dangerous levels. Given the technical, social and ecological barriers that need to be overcome for it to work at the required scale, it seems wiser to not bank on BECCS or any other NET coming to the rescue. In fact, it is arguably totally unethical to bank on them, or to build a *reassuring* 'Net Zero by 2050' story that relies on them - but barely mentions them.

It is perhaps better to apply the precautionary principle, assume that NETs will only make a small contribution in the #RaceToZero and plot pathways to 1.5°C / 2°C that do not rely on them. However, without a reliance on NETs, the action

Reassuring stories about the planet and its future are nothing new - this book is reporting on just one of them. In early 2021, filmmaker Ali Tabrizi released Seaspiracy, a feature length Netflix documentary film on the devastating impacts of the fishing industry[159]. It had echoes of Cowspiracy, the 2014 exposé of industrial livestock farming[160]. Neither film held back, and both caused controversy - as films that tear reassuring stories apart should. It is partly because of the naivety that the two films got criticised for, that they are so powerful. They slap their audiences round the face with new angles on the farming and fishing stories that are so often concealed or misrepresented. It is a blunt method and inevitably imperfect, but whether or not audiences agree with the conclusions, they at least have to think about them.

Films like Seaspiracy create what educators call a disorientating dilemma, they force their audiences into an uncomfortable cognitive and emotional state that they *have* to respond to. Some will adopt the 'oh dear' response and forget it like a bad dream, but many others will set off on a learning journey to find out more about the issue and, consequently, themselves. These films are worth making for that reason; they give the reassuring stories a good shake. They ask if revered environmental broadcasters, campaigners, and NGOs should be as trusted as they are, and they have the power to dramatically re-frame the debate around the issues they cover. It is only a matter of time before another film comes along that smashes the reassuring stories about NETs and Net Zero to pieces.

required to stay below 2°C is even more urgent, the situation even more stark. Kevin Anderson and Isak Stoddard estimate that to stay within its carbon budget the UK would need to ramp up its mitigation effort incredibly fast. It would need to cut greenhouse gas emissions by 10% per year by 2025, 20% per year by 2030 and achieve a 'Real Zero' (not Net Zero) energy system by around 2035[161].

Achieving even half of these cuts would be extraordinary, full decarbonisation by 2035 would require a mammoth effort. Anderson equates the task ahead to something akin to FDR's 1930s New Deal plus the 1948 Marshall plan[162]. Or, to be hyperbolic for a moment, a *Green New Deal Marshall Plan - on steroids*. The disruption and reorganisation of society would be colossal. This is what the Transformative Adaptation movement (see chapter 8) is talking about. It is not a little bit of tweaking, it is wholesale change - a complete revision of *business-as-usual*.

So that's 'NETs' and the 'BECCS' shaped holes in them. The reassuring story glosses over a whole lot more - 'direct air capture', 'solar radiation management', 'ocean fertilisation' and many other geoengineering ideas pop up in the media to tell us that technology will come to the rescue. But what hasn't yet emerged is something that is plausible, scalable, and – most importantly – just. But the *reassuring story* lives on, and it even has a part two.

## The reassuring story (part two)

As climate change worsens and part one of the *reassuring story* starts to fall apart, the situation people find themselves in will become more frightening. They will need reassuring all over again, this will open the door for reassuring story part two. It will talk in familiarly comforting tones, but this time about the adaptation programmes that powerful actors are putting in place to keep us safe as climate change hits. In this story, the solutions will once again be framed as scientific and big-tech, adaptation will be presented as a top-down process that is done for us, for our own good, by leaders who are 'world-leading' at it. Again, the emphasis will be on incremental change and minimal disruption to the status quo. It is a troubling prospect. So, just as 'reassuring story part one' needs to be challenged, so too does 'part two', because it is already coming to the fore. The task is to reframe adaptation in the consciousness of the climate movement and the general population.

---

In 2020, Rachel Harcourt and colleagues[163] revealed that by far the most dominant adaptation story told by UK newspapers was this one: *There has been a flood. It has caused a lot of damage. The Government is to blame. We need more flood barriers. The Government needs to build them.* There is sometimes an addendum to this story, it urges you to check your home insurance coverage and your local authority's flood prevention plan. You might also find a story or two about a farmer or an animal that is trying to adapt, but that's about it. Headline grabbing stuff it is not, especially if you don't live on an exposed floodplain, which most people don't.

Even within the environmental sector, adaptation stories struggle to cut through and when they do, they are either attacked as being defeatist, or dismissed as something environmentalists shouldn't (right now) concern themselves with. In 2020, The Glacier Trust found that only 0.82% of articles written by the UK's five biggest environmental organisations are focused on climate change adaptation[164]. It does get the odd passing (not always positive) mention, but articles focussed specifically on adaptation remain a very rare find in the environmentalist world.

For those who do want to engage with adaptation, there are a few niche email lists you can sign up to[165] and – if Bill Gates sponsored initiatives are your thing – there is the newly formed Global Center on Adaptation[166]. Other than that, you can (a) read sector specific case studies about the adaptation efforts of various governments and institutions; (b) trawl through a few lengthy academic papers or PDF downloads with headlines like *Climate resilient development pathways - an ecosystem-based adaptation programme for England*; or (c) delve into dense textbooks with equally procedural titles. Anyone for *Adaptation Policy Frameworks for Climate Change?* It is no wonder climate change adaptation has an image problem.

But as tempting (and easy) as it is to just ignore adaptation, it is a risky strategy. As this book has shown, adaptation *is* happening and *will* go on happening, and at an ever-greater scale – whether we like it or not. This is not necessarily a good thing, not for everyone, especially if the responsibility for adaptation is left in the hands of central Governments, large NGOs, and big businesses that are, by nature, resistant to anything truly transformative. Their aim will be to soothe us with a reassuring story about adaptation being – like mitigation – on track, under control, and safe in their *very capable hands*.

If you accept this, you are accepting that decisions will be made for you, not with you, nor by you. And if you are prepared to allow this, you must be prepared to accept 'solutions' that might not be ones you would have wanted if you had been asked - or, if you had spoken up. Things *might* work out alright if we – as citizens – go along with the plans being made... but they might not. The adaptation 'solutions' we are given, might end up being as flawed and incremental as the mitigation 'solutions' that reassuring story part one is selling us. Your, and our, worst fears about adaptation and *mal*adaptation might start coming true.

Rachel Harcourt's paper, 'What Adaptation Stories are UK Newspapers Telling?' issues us with a warning. She and her colleagues showed that a specific story about adaptation is starting to take root. It makes the need to speak up not only important, but urgent. Newspapers are framing adaptation as a technical process, done *by* the powerful *to* the rest of us, to protect society as it currently is. If this story sticks – and it might if it is left unchecked – it will shape what society understands adaptation to be and what (and who) it is understood to be for. This will, in turn, shape what it is, and who benefits. Why? Because, as anyone who works in PR will tell you, stories about the future, if told often enough, can start to come true - especially when rich and powerful people are the ones telling it.

If we are told that adaptation is something that is done to us – and if we don't challenge that story or hear a different one – it will *become* something that is done to us. Adaptation won't be *by*

## THE REASSURING STORY

the people, *for* the people, it will be folded into the wider project of keeping today's jagged socio-economic structures firmly intact. If we don't want adaptation to be done to us, and don't want those structures to remain intact, we need to mainstream new and engaging adaptation stories. We need to improve adaptation's image and attach it to stories of transformation. Only by acting together now to construct and tell these new stories, repeatedly, to the world's changemakers, can we win the battle to shape the way adaptation is understood in the public consciousness.

## 12. Adaptation is unavoidable, but *mal*adaptation is not

To some in the climate movement, adaptation is the 'A' word[167]. They don't like talking about it, and they don't like other people talking about it. This is especially true when adaptation and mitigation are placed in opposition to each other and presented as an either/or choice that one has to pick a side on. Thankfully these divisions and attitudes aren't as widespread as they once were, but they still linger. There are people who aren't necessarily anti adaptation, but they are not vocally pro either. They remain agnostic, preferring to remain silent, citing scepticism about the harm a positive narrative on adaptation might do to efforts on mitigation. Some go further and attack the motives of those who advocate for adaptation. The most cynical frame adaptation narrowly, dismissing it as something that only selfish actors do.

Scepticism about adaptation is understandable, especially given how adaptation has thus far been framed. When understood as a top-down technocratic effort, a selfish pursuit, or the plaything of a billionaire who wants to play God, adaptation looks ugly. The other commonly used adaptation frames like 'mitigation', 'security', 'technoscientific' and 'ecological'[168] have their strengths, but none of them have managed to cut through and position adaptation as something that has universal appeal. In this context, the ongoing reluctance to talk about adaptation makes sense. However, not talking about adaptation and hoping it will go away is not the way forward. Adaptation is going to happen (is happening) and stories *will* be told about it. It is better to engage with it now, while it is still a relatively young topic, so that it can be shaped and framed in ways that complement other critical agendas.

To cut through the scepticism, the *reassuring stories*, the apathy, and the fear, we can re-frame adaptation as something citizen-led, something *just*, and something that supports broader goals of social and economic transformation. If we can get *that* story of adaptation to stick, it will be *that* story that starts to shape how adaptation is thought of. It will then be more likely to be practiced in ways that are ecologically sensitive and socially just. In that form, adaptation can supplement mitigation and transformation efforts, not work against them, both in the real world and in our minds.

---

Given the turbulence the world is currently experiencing, there are arguments to be made that this is the best time to do something drastic and transformative - an appetite for change hangs in the air. Alternatively, it can be argued that this is the absolute worst time to abandon the reassuring story and expose the true enormity of what lies ahead - who in their right mind would want to do anything other than try to re-create a sense of calm and order?

Ultimately though, climate change is here. Global average temperatures have increased by at least 1.2°C and the UN backed plans being made to prevent it getting at least one degree hotter appear critically flawed. The whistle will soon be blown, the *reassuring stories* will lose their ability to mollify and pacify, and something new will force its way through. The frustration is mounting, when Greta Thunberg spoke at the UN in New York, it was outrage she was expressing, it wasn't an act[169]. She is the superstar environmentalist, but she is far from alone. The fury is growing and spreading as more and more people wake up to the yawning gaps between target setting rhetoric, and real world action. If you're not outraged already by what is happening, you will be soon. But don't let that anger and outrage fester, harness it, because once the reassuring stories have fallen apart - and they will - change will happen; generation Greta is uniquely equipped to ensure that it does.

Young people have access to more information than any generation ever before and, more importantly, they have the ability to analyse and synthesise what they are learning. They are critical, empathic, and systemic thinkers in ways previous generations weren't. This 'ecological intelligence'[170] enables them to join up the dots between issues that have for so long been presented as separate. And, as they discover the links and the intersections, their outrage, anger and motivations grow. They understand that gender inequality, racial discrimination, environmental degradation, and social injustice aren't isolated problems that are fixable within an otherwise fair and functioning system. They see that the system itself is malfunctioning. They therefore reject invitations to take actions that feel tokenistic, infantilising, or siloed, and call instead for deeper, more profound change. Today's young people are thinking systemically about problems, and therefore systemically about solutions. It is inspiring to witness and a source of great hope.

Climate change is here, and further disruption is unavoidable. Adaptation is therefore unavoidable too, but *mal*adaptation isn't.

*Great Adaptations* are possible, societal transformation is possible. So, as well as being outraged, allow yourself to be fired up. Spend time with those who are interrogating the deeply held beliefs about economics and human nature that are propping up a malfunctioning system. Work with them to re-shape those beliefs at a societal level, so that new forms of politics and economics can emerge. Different and better visions of the future are out there, others can be imagined, systemic change is possible, new futures can be built.

## Postscript

The effects of the warming that has already happened are not benign, lives and livelihoods – all over the world – have already been upended. But animals, plants and people aren't simply letting climate change happen to them, they are adapting. Some adaptations are carefully planned and resourced, others emphatically aren't. They all have consequences. Every adaptation covered in this book – and the many others that didn't make the cut – tells us something about the adaptations to come. It is crucial we learn from the early adapters and from their successes and failures, for to ignore them is to ignore our own future.

*Great Adaptations* is a book, but also a campaigning resource. Supporters of The Glacier Trust are purchasing copies of this book for themselves, as well as to pass onto key figures and influencers in the environmental movement[171]. We hope you have enjoyed it and feel moved to engage with adaptation both personally and professionally.

Finally, we ask you to join us. As chapter 11 laid out, climate change adaptation has an image problem: current narratives around it are largely unhelpful and prevent it from cutting through in the environmental movement (let alone with the public more broadly). The Glacier Trust is trying to frame adaptation as a positive and transformative process grounded in the principles of social justice and ecological enhancement. This book has tried to do this, but more work needs to be done by us and others to refine the way adaptation stories are told. So, in telling and re-telling the adaptation stories you read in this book and elsewhere, think about *how* you tell them too: the values, beliefs, paradigms, and worldviews you are activating and reinforcing matter. Being aware of this is how, together, we can transform the way adaptation is thought about and done. We need stories of adaptations that are citizen-led, participatory, and flexible enough to withstand a potentially *much*-changed climate.

To ensure that these are the stories that stick, and the stories that shape the future of adaptation, the storytellers need inspiring examples of adaptation strategies that are proven and scalable. They will also need warnings from history of how *not* to adapt, and a language of adaptation to draw upon. This book is a start, but more books are needed... and films, and poems, and podcasts, and songs. This is the work of the adaptation advocate, you can become one.

For more information on the work of The Glacier Trust please visit our website: **www.theglaciertrust.org**. You can also find us on Facebook, Twitter and Instagram.

Any profits made from the sale of this book will be used to support the work of The Glacier Trust. We will use it to enable climate change adaptation in Nepal.

If you would like to make a larger contribution to our work, please visit **donate.theglaciertrust.org/donate** to set up a one-off or monthly donation. Alternatively, if you are in the UK, you can **text GREAT to 70085** to make a £5 donation[1].

---

Thank you.

[1] Texts cost £5 plus one standard rate message and you'll be opting in to hear more about our work and fundraising via telephone and SMS. If you'd like to give £5 but do not wish to receive marketing communications, text GREATNOINFO to 70085.

# Reference list

This list can also be found on The Glacier Trust website: **theglaciertrust.org/great-adaptations/references**. It includes clickable links to all articles and publications referenced.

[1] Phillips, M., Richards, C., Bunk, P. (2020) *We Need To Talk About Adaptation 2020*, The Glacier Trust

[2] Moore, M., Phillips, M. (2019) *Coffee. Climate. Community.* The Glacier Trust

[3] Arkbound Foundation (2021) *Climate Adaptation: Accounts of Resilience, Self-Sufficiency and Systems Change*, Arkbound

[4] UNFCCC (2015) *The Paris Agreement*, UNFCCC

[5] UN Environment (2020) *Emissions Gap Report 2020*, UNEP

[6] Read. R. (2021) *Parents for a future - How loving our children can prevent climate catastrophe*, UEA Publishing Project

[7] Liu, P.R., Raftery, A.E. (2021) *Country-based rate of emissions reductions should increase by 80% beyond nationally determined contributions to meet the 2°C target.* Communications Earth and Environment, 2 (29)

[8] WMO (2021) *The State of the Global Climate 2020*, World Meteorological Organization

[9] IPCC (2018) *Global Warming of 1.5°C. An IPCC Special Report on the impacts of global warming of 1.5°C above pre-industrial levels and related global greenhouse gas emission pathways, in the context of strengthening the global response to the threat of climate change, sustainable development, and efforts to eradicate poverty*, [Masson-Delmotte, V., P. Zhai, H.-O. Pörtner, D. Roberts, J. Skea, P.R. Shukla, A. Pirani, W. Moufouma-Okia, C. Péan, R. Pidcock, S. Connors, J.B.R. Matthews, Y. Chen, X. Zhou, M.I. Gomis, E. Lonnoy, T. Maycock, M. Tignor, T. Waterfield (eds.)]

[10] Hausfather, Z., Peters, G. (2020) *Comment: Emissions - the 'business as usual' story is misleading*, Nature Climate Change, vol 557, 618-620

[11] McSweeney, R. (2020) *Explainer: Nine 'tipping points' that could be triggered by climate change*, Carbon Brief

[12] Betts, R. (2018) *Hothouse Earth: here's what the science actually does – and doesn't – say*, The Conversation

[13] Lenton. T.M., et. al (2019) *Climate tipping points — too risky to bet against*, Nature

[14] Ilyas-Jarrett, S. (2020) *Why don't we take climate change seriously? Racism is the answer*, Open Democracy

[15] Lammy, D. (2020) *Climate justice can't happen without racial justice*, TED talk

[16] Gardiner, B. (2020) *Unequal Impact: The Deep Links Between Racism and Climate Change*, Yale Environment 360

[17] UN Environment (2021) Adaptation Gap Report 2020, UNEP

[18] Hickel, J. (2017) *The Divide*, Penguin Random House
[19] Haraway, D. (2016) *Staying With The Trouble*, Duke University Press
[20] Pereira, I. (2017) *Sandy stories: Firsthand accounts of surviving the storm*, amNY
[21] Koslov, L. (2019) *Avoiding Climate Change: "Agnostic Adaptation" and the Politics of Public Silence*, Annals of the American Association of Geographers, 109 (2)
[22] Pilkey, O. (2012) *We Need to Retreat From the Beach*, The New York Times
[23] Farand, C. (2021) *Cyclone Eloise shatters Mozambique's progress to recover from 2019 storms*, Climate Change News
[24] Fischer, K.K. (2015) *"Agnostic Adaptation."* In *'A Response to the IPCC Fifth Assessment,'* by Adams-Schoen, S. et al. Environmental Law Reporter 45, no. 1: 10027 – 48
[25] Members of District of Warwick people's inquiry on climate change (2021) *Jury Statement*, Warwick District Council
[26] Ritchie, H., Roser, M. (2019) *Urbanization*, Our world in data
[27] UN (2018) *68% of the world population projected to live in urban areas by 2050, says UN*, United Nations
[28] O'Neill, C. (2020) *Letter to Prime Minister Johnson, February 3rd 2020*
[29] BBC News (2021) *Climate change is a threat to our security - Boris Johnson*, BBC News
[30] Scottish Government (2020) *Public bodies climate change duties: putting them into practice, guidance required by part four of the Climate Change (Scotland) Act 2009*
[31] Sniffer (2021) *Vision, mission and values*
[32] Climate Ready Clyde (2021) *Our Adaptation Strategy and Action Plan*
[33] Climate Ready Clyde (2018) *Towards a climate ready Clyde; climate risks and opportunities for the Glasgow City Region*
[34] Caroline (2020) *5 Best Cooling Mats for Dogs*, Rangers Dog
[35] BBC News (2019) *UK heatwave: Met Office confirms record temperature in Cambridge*, BBC News
[36] Knapman, H. (2019) *SWEATING IT Brits panic in heatwave and strip shops including Asda and Tesco of fans, ice creams and cooling mats for pets*, The Sun
[37] Designrhome.com (2019) *10 Mist Cooling Fans That Bring Relief In Hot Temperatures*
[38] Bakar, F. (2019) *B&M is selling sun loungers for dogs so you can both chill in the garden*, Metro
[39] Symester, C. (2019) *Currys PC World slash price of fans as UK shoppers try to combat scorching weather*, The Daily Mirror
[40] Clement, M. (2019) *I followed the advice for Paris's hottest day - it didn't help*, The Guardian
[41] Dalton, J. (2019) *Qatar now so hot it has started air-conditioning the outdoors*, The Independent
[42] Orton, K. (2012) *The desert of the unreal*, Dazed Digital
[43] Western Sydney University (2020) *New Resource by Institute Researchers Provides Advice on How to Prepare for Heat*
[44] Earth IQ (2019) *How can we stay cool without contributing to climate change?* Facebook
[45] Lopes, A. et.al. (2019) *Infrastructures of Care: Opening up 'Home' as Commons in a Hot City*, Human Ecology Review, 24 (2)

# REFERENCE LIST

[46] Wood, S. (2004) *Every skier should have some more snow*, The Independent

[47] Clavarino, T. (2019) *The Guardian picture essay: Seduced and abandoned: tourism and climate change in the Alps*, The Guardian

[48] Willshere, K. (2019) *French ski resort moves snow with helicopter in order to stay open*, The Guardian

[49] Knox, P. (2019) *SNOW WAY Russia using FAKE snow on Moscow streets as it records warmest winter on record*, The Sun

[50] Morales-Castilla, I., et.al. (2020) Diversity buffers winegrowing regions from climate change losses, Proceedings of the National Academy of Sciences, 117 (6)

[51] French, P. (2019) *Champagne Taittinger expands English vineyard ahead of first harvest*, The Drinks Business

[52] NATO (2014) *Environment - NATO's stake*, NATO

[53] Givetash, L. (2019) *Militaries go green, rethink operations in face of climate change*, NBC News

[54] Neimark, B. et. al. (2019) *US military is a bigger polluter than as many as 140 countries - shrinking this war machine is a must*, The Conversation

[55] NATO (2014) *Environment - NATO's stake*, NATO

[56] The Economist (2019) *How climate change can fuel wars*

[57] Stoltenberg, J. (2019) *Speech by NATO Secretary General Jens Stoltenberg at the Institute for Regional Security and the Australian National University's Strategic and Defence Studies Centre, Canberra*, NATO

[58] Groeskamp, S., Kjellsson, J. (2020) *NEED The Northern European Enclosure Dam for if Climate Change Mitigation Fails*, BAMS Article, American Meteorological Society (July, 2020)

[59] Henley, J., Evans, A. (2020) *Giant dams enclosing North Sea could protect millions from rising waters*, The Guardian

[60] Friedman, M. (1982) *Capitalism and Freedom*, Phoenix Books

[61] U.S. Army Corps of Engineers, Norfolk District (2020) *Miami-Dade Back Bay Coastal Storm Risk Management Draft Integrated Feasibility Report and Programmatic Environmental Impact Statement Miami-Dade County, Florida*

[62] Harris, A. (2020) *Feds have $4.6 billion plan to protect Miami-Dade from hurricanes: walls and elevation*, Miami Herald

[63] UN Global Compact (2015) *The Business Case For Responsible Corporate Adaptation*, United Nations

[64] Holthaus, E. (2021) *The Phoenix*

[65] Holthaus, E. (2020) *Big disasters make headlines. But the most dangerous part of climate change is that you barely notice it's happening*, The Conversation

[66] Magnan, A. (2014) *Avoiding maladaptation to climate change: towards guiding principles*, SAPIENS, 7 (1)

[67] The Biomimicry Institute (2020) *What is Biomimicry?*

[68] Peacock, E., Sonsthagen, S.A., Obbard, M.E., Boltunov, A., Regehr, E.V., et al. (2015) *Implications of the Circumpolar Genetic Structure of Polar Bears for Their Conservation in a Rapidly Warming Arctic*, PLOS ONE 10(8)

[69] Peng, G. et.al., (2020) *What Do Global Climate Models Tell Us about Future Arctic Sea Ice Coverage Changes?*, Climate

[70] Watts, J. (2019) *What polar bears in a Russian apartment block reveal about the climate crisis*, The Guardian

[71] Letzer, R. (2020) *Australian Hunters to Kill 10,000 Feral Camels from Helicopters Amid Worsening Drought*, Live Science

[72] Betz, B. (2020) *In Australia, more than 5,000 feral camels killed in mass cull*, Fox News

[73] Oregon State University (2013) *The sounds of science: Melting of iceberg creates surprising ocean din*, Phys org

[74] Wallace-Wells, D. (2019) *The Uninhabitable Earth*, Allen Lane

[75] Cooke, R.S.C., Eigenbrod, F., Bates, A.E. (2019) Projected losses of global mammal and bird ecological strategies, Nature Communications, 10 (2279)

[76] Milton, N. (2019) *Adders now active all year with warmer UK weather*, The Guardian

[77] Earth Institute Columbia University (2021) *Renee Cho Author at State of the Planet*

[78] Cho, R. (2018) What Helps Animals Adapt (or Not) to Climate Change? Earth Institute, Columbia University

[79] Cho, R. (2018) What Helps Animals Adapt (or Not) to Climate Change?, State of the Planet, Earth Institute Columbia University

[80] Bregman, R. (2020) *Humankind: A Hopeful History*, Bloomsbury Publishing

[81] Hare, B., Wood, V. (2020) *Survival of the Friendliest - Understanding our Origins and Rediscovering Our Common Humanity*, Penguin Random House

[82] Dar Si Hmad (2021) *Vision & Mission*

[83] Warwick District Council (2020) *Warwick District's Climate Emergency Action Programme*

[84] Butler, P. (2020) *Warwick asks voters to back radical council tax rise for climate action*, The Guardian

[85] Warwick District Council (2021) *People's inquiry sessions*

[86] Climate Action Now (2021) Warwick District Council

[87] Warwick District Council (2020) *Warwick District's Climate Emergency Action Programme*

[88] Climate Change Committee (2017) *UK Climate Change Risk Assessment 2017 Evidence Report*

[89] Sobczak-Szel, K., Fekih, N. (2020) *Migration as one of several adaptation strategies for environmental limitations in Tunisia: evidence from El Faouar*, Comparative Migration Studies, 8 (8)

[90] The Climate and Migration Coalition: climatemigration.org.uk

[91] Hulme, M. (2019) *Climate Emergency Politics Is Dangerous*, Issues in Science and Technology

[92] HM Government (2019) *Green GB & NI - Industrial strategy*

[93] Committee on Climate Change (2019) *Net Zero - The UK's contribution to stopping global warming*

[94] UN News (2019) *UN emissions report: World on course for more than 3 degree spike, even if climate commitments are met*, UN News

[95] Simms, A. (2017) *'A cat in hell's chance' - why we're losing the battle to keep global warming below 2C*, The Guardian

[96] Climate Action Tracker (2020) *Global update: Paris Agreement Turning Point*

# REFERENCE LIST

[97] Knorr, W. et.al. (2020) *Letters: After coronavirus, focus on the climate emergency*, The Guardian

[98] Mann, G., Wainwright, J. (2017) *Climate Leviathan: A Political Theory of Our Planetary Future*, Verso

[99] Dobson, J. (2020) *Billionaire Bunker Owners Are Preparing For The Ultimate Underground Escape*, Forbes

[100] The Seasteading Institute (2019) *Reimagining Civilization with Floating Cities*

[101] Milman, O., Rushe, D. (2021) *The latest must-have among US billionaires? A plan to end the climate crisis*, The Guardian

[102] Schipper, E.L.F. (2020) *Maladaptation: When adaptation to climate change goes very wrong.* One Earth, 3(4)

[103] Bendell, J., Read, R. (2021) *Deep Adaptation Navigating the Realities of Climate Chaos*, Polity Press

[104] Bendell, J. (2020) *Deep Adaptation: A Map for Navigating Climate Tragedy*, 2nd Edition, IFLAS Occasional Paper 2

[105] Nicholas, T., Hall, G., Schmidt, C. (2020) *The faulty science, doomism, and flawed conclusions of Deep Adaptation*, Open Democracy

[106] Read, R., Eastoe, J. (2021) *The Need for a 'Moderate Flank' in climate activism*, Byline Times

[107] Anonymous (2017) *This Civilisation is Finished...*, Green Talk

[108] Read, R. (2018) *This civilisation is finished: so what is to be done?* Speech to Shed a light at Churchill College, University of Cambridge on 7 November 2018, YouTube

[109] Read, R. (2019) *This civilisation is finished: Conversations on the end of Empire - and what lies beyond.* Simplicity Institute

[110] Read, R. (2021) *Transformative Adaptation*, Permaculture Magazine, Vol. 10, Spring 2021

[111] Roser, M., Ortiz-Ospina, E. (2019) *Global Extreme Poverty*, Our World in Data

[112] Dorling, D. (2020) *Slowdown: The End of the Great Acceleration — and Why It's Good for the Planet, the Economy, and our Lives*, Yale University Press

[113] Rao, K. (2016) *Amitav Ghosh 'climate change is like death, no one wants to talk about it'*, The Guardian

[114] Gergish, J. (2020) *The great unravelling: 'I never thought I'd live to see the horror of planetary collapse'*, The Guardian

[115] Curtis, A. (2016) *Hypernormalisation*, BBC iPlayer

[116] Curtis, A. (2009) *Oh dearism*, BBC iPlayer

[117] Sawin, E. (2019) *Obvious to you, but invisible to me*, The Glacier Trust

[118] Yurchak, A. (2005) *Everything was Forever, Until it was No More: The Last Soviet Generation*, Princeton University Press.

[119] Klein, N. (2007) *The Shock Doctrine: The Rise of Disaster Capitalism*, Allen Lane

[120] Fischetti, M. (2015) *Climate Change Hastened Syria's Civil War*, Scientific American

[121] Internationalist Commune of Rojava (2018) *Make Rojava Green Again*, Dog Section Press

[122] Ross, C. (2017) *Accidental Anarchist*, Hopscotch Films

[123] Bookchin, D. (2019) *Report from Rojava: What the West Owes its Best Ally Against ISIS*, The New York Review

[124] Marvel, K. (2018) *We Need Courage, Not Hope, to Face Climate Change*, On Being

[125] Anthony, A. (2017) *Ex-diplomat Carne Ross: the case for anarchism*, The Guardian

[126] Crompton, T., Sanderson, R., Prentice, M., Weinstein, N., Smith, O., Kasser, T.(2016) *Perceptions Matter - The Common Cause UK Values Survey*, Common Cause Foundation

[127] Parker, N., Phillips, M. (2021) *United in Compassion: Bringing young people together to create a better world*, Global Action Plan

[128] Willis, R. (2020) *Too Hot to Handle?: The Democratic Challenge of Climate Change*, Bristol University Press

[129] O'Brien, K. (2018) *The three spheres of transformation*, cCHange

[130] Cottam, H. (2020) *Welfare 5.0: Why we need a social revolution and how to make it happen*, UCL

[131] Bregman, R. (2020) *Humankind: A Hopeful History*, Bloomsbury

[132] Hare, B., Woods, V. (2020) *Survival of the Friendliest: Understanding Our Origins and Rediscovering Our Common Humanity*, Oneworld Publications

[133] Common Cause Foundation: *valuesandframes.org*

[134] Lifeworlds Learning: *lifeworldslearning.co.uk*

[135] New Citizenship Project: *newcitizenship.org.uk*

[136] Global Action Plan: *globalactionplan.org.uk*

[137] Hickel, J. (2020) *Less is More: How Degrowth Will Save the World*, Penguin

[138] Raworth, K. (2018) *Doughnut Economics Seven Ways to Think Like a 21St-Century Economist*, Random House

[139] Gonick, L., Kasser, T. (2018) *Hyper-Capitalism*, Scribe

[140] Bookchin, M. (2015) *The Next Revolution: Popular Assemblies and the Promise of Direct Democracy*, Verso

[141] Ross, C. (2011) *The Leaderless Revolution: How Ordinary People Will Take Power and Change Politics in the 21st Century*, Simon & Schuster UK

[142] Dorling, D., Koljonen, A. (2020) *Finntopia - What We Can Learn From the World's Happiest Country*, Agenda Publishing

[143] Macy, J. (2018) *It Looks Bleak. Big Deal, It Looks Bleak.*, Ecobuddism

[144] Aronhoff, K. (2019) *Things Are Bleak!*, The Nation

[145] Global Campaign to Demand Climate Justice (2020) *NOT ZERO: How 'net zero' targets disguise climate inaction*, Joint technical briefing by ActionAid, Corporate Accountability, Friends of the Earth International, Global Campaign to Demand Climate Justice, Third World Network, and What Next?

[146] Jackson, T. (2019) *2050 is too late - we must drastically cut emissions much sooner*, The Conversation

[147] Lucas, C. (2019) *Theresa May's net-zero emissions target is a lot less impressive than it looks*, The Guardian

[148] Anderson, K., Broderick, J.F., Stoddard, I. (2020) *A factor of two: how the mitigation plans of 'climate progressive' nations fall far short of Paris-compliant pathways*, Climate Policy, 20 (10)

# REFERENCE LIST

[149] Weston, P. (2019) *Zero carbon 2050 pledge is too slow to address catastrophic climate change, campaigners warn*, The Independent

[150] Harvey, F. (2020) *Ministers doing little towards 2050 emissions target, say top scientists*, The Guardian

[151] Kirkpatrick, B. (1996) *Brewer's concise dictionary of phrase and fable*, Cassell

[152] Climate Change Committee (2021) *Independent Assessment of UK Climate Risk - Advice to Government For the UK's third Climate Change Risk Assessment (CCRA3)*

[153] DEFRA (2018) *The National Adaptation Programme and the Third Strategy for Climate Adaptation Reporting - Making the country resilient to a changing climate*

[154] ADM (2021) *Ethanol*, ADM

[155] Brack, D., King, R. (2020) *Research Paper: Net Zero and Beyond: What Role for Bioenergy with Carbon Capture and Storage?* Chatham House

[156] Peters, G. (2017) *Does the carbon budget mean the end of fossil fuels?* Energi og Klima

[157] Hickel, J. (2020) *Less is More: How Degrowth Will Save the World*, Penguin

[158] United Nations (2018) *The World's Cities in 2018*, The United Nations

[159] Tabrizi, A. (2021) *Seaspiracy*, Netflix

[160] Andersen, K., Kuhn. K. (2014) *Cowspiracy: The Sustainability Secret*

[161] Anderson, K., Stoddard, I. (2020) *Beyond a climate of comfortable ignorance*, The Ecologist

[162] Anderson, K. (2020) *Revisiting Tressell's philanthropists in the light of the Covid-19 and Climate emergencies*, YouTube

[163] Harcourt, R., Bruine de Bruin, W., Dessai, S., Taylor, A. (2020) *What Adaptation Stories are UK Newspapers Telling? A Narrative Analysis*, Environmental Communication, 14 (8)

[164] Phillips, M., Richards, C., Bunk, P. (2020) *We Need To Talk About Adaptation 2020*, The Glacier Trust

[165] Sign up to the Adaptation Scotland e-newsletter here: *adaptationscotland.org.uk/news-events*; the WeAdapt e-newsletter here: *weadapt.org*; and the Cultural Adaptations e-newsletter here: *culturaladaptations.com/news*

[166] Global Center on Adaptation: *gca.org*

[167] Revkin, A. (2019) *Once derided, ways of adapting to climate change are gaining steam*, National Geographic

[168] Richards, C. (2020) *Framing Adaptation*, The Glacier Trust

[169] Thunberg, G. (2019) *Transcript: Greta Thunberg's Speech At The U.N. Climate Action Summit*, NPR

[170] Sterling, S. (2009) *Ecological Intelligence: viewing the world relationally*, [In] Stibbe, A. (2009) *The Handbook of Sustainability Literacy*, Green Books

[171] The Glacier Trust (2021) *Great Adaptations*, The Glacier Trust: *theglaciertrust.org/great-adaptations*

# Photo credits

"Portrait of Morgan Phillips for #wearestillhere project" by Alex Basaraba

"Black Lives Matter march down I-35" by Fibonacci Blue is licensed under CC BY 2.0

"Mother Nature Wants Her Land Back" by Liz Koslov

"Sun loungers for pets" by Amanda Collins

"The School of Design and Environment building at the National University of Singapore" by Rory Gardiner

"Snow machine" by CLS Rob is licensed under CC BY-SA 2.0

"Vinyard" by Ignacio Morales-Castilla

"180925-Z-XH297-1108" [Flooded property in Miami, Florida, USA, in the aftermath of Hurricane Florence] U.S. Army National Guard Photo by Staff Sgt. Jorge Intriago is marked with CC0 1.0

"Coffee Harvest in Deusa, Solukhumbu, Nepal" by Meleah Moore, The Glacier Trust

"Pulping coffee cherries in Deusa, Solukhumbu, Nepal" by Meleah Moore, The Glacier Trust

"Chott El Djerid" by David Stanley is licensed under CC BY 2.0

"Anticolonial March Berlin 2019 Rojava" by Leonhard Lenz is marked with CC0 1.0